T0311762

Cambridge Elements ≡

Elements in Bioethics and Neuroethics
edited by
Thomasine Kushner
California Pacific Medical Center, San Francisco

CAPACITY, INFORMED CONSENT, AND THIRD-PARTY DECISION-MAKING

Jacob M. Appel
Icahn School of Medicine at Mount Sinai

CAMBRIDGE
UNIVERSITY PRESS

CAMBRIDGE
UNIVERSITY PRESS

Shaftesbury Road, Cambridge CB2 8EA, United Kingdom

One Liberty Plaza, 20th Floor, New York, NY 10006, USA

477 Williamstown Road, Port Melbourne, VIC 3207, Australia

314–321, 3rd Floor, Plot 3, Splendor Forum, Jasola District Centre, New Delhi – 110025, India

103 Penang Road, #05–06/07, Visioncrest Commercial, Singapore 238467

Cambridge University Press is part of Cambridge University Press & Assessment, a department of the University of Cambridge.

We share the University's mission to contribute to society through the pursuit of education, learning and research at the highest international levels of excellence.

www.cambridge.org
Information on this title: www.cambridge.org/9781009570091

DOI: 10.1017/9781009570060

First published 2024

A catalogue record for this publication is available from the British Library.

ISBN 978-1-009-57009-1 Hardback
ISBN 978-1-009-57008-4 Paperback
ISSN 2752-3934 (online)
ISSN 2752-3926 (print)

Capacity, Informed Consent, and Third-Party Decision-Making

Elements in Bioethics and Neuroethics

DOI: 10.1017/9781009570060
First published online: July 2024

Jacob M. Appel
Icahn School of Medicine at Mount Sinai
Author for correspondence: Jacob M. Appel, jacobmappel@gmail.com

Abstract: This Element examines three related topics in the field of bioethics that arise frequently both in clinical care and in medico-legal settings: *capacity, informed consent, and third-party decision-making.* All three of these subjects have been shaped significantly by the shift from the paternalistic models of care that dominated medicine in the United States, Canada, and Great Britain prior to the 1960s to the present models that privilege patient autonomy. Each section traces the history of one of these topics and then explores the major ethics issues that arise as these issues are addressed in contemporary clinical practice, paying particular attention to the role that structural factors such as bias and social capital play in their use. In addition, the volume also discusses recent innovations and proposals for reform that may shape these subjects in the future in response both to technological advances and changes in societal priorities.

Keywords: bioethics, capacity, informed consent, surrogates, competence

ISBNs: 9781009570091 (HB), 9781009570084 (PB), 9781009570060 (OC)
ISSNs: 2752-3934 (online), 2752-3926 (print)

Contents

1 Introduction

This Element discusses three complex and interrelated subjects at the nexus of bioethics and law: capacity assessment, informed consent, and the role of advance directives and third-party representatives in healthcare decision-making. Although physicians and courts have grappled with these issues since the nineteenth century, rapid changes in American healthcare ethics and the emergence of an autonomy-driven medical culture over the past six decades have seen a radical transformation of the legal and philosophical landscape in these areas. A medico-legal authority arriving from the 1960s to round with a clinical ethics team at a hospital in the United States or Great Britain today would likely find himself in unrecognizable territory. Substantial progress has been made toward empowering patients and ensuring justice. However, that does not mean that these fields remain static or that more challenges do not await. Revolutionary technological advances continue to reshape both diagnostic tools and clinical remedies. Recent steps toward more inclusion and equity serve as a reminder that far more needs to be done to ensure that every patient is afforded the same protections and opportunities. In all likelihood, a present-day medico-legal expert arriving in the year 2084 would find the contents of this short volume entirely alien. That is as it should be. Bioethics is a dynamic field. In the interim, these pages offer a brief primer for those who wish to explore several of the most intriguing, if challenging, fields in contemporary medical law.

This is not a traditional textbook. Nor is it intended to be comprehensive. Rather, the volume explores the major legal and ethical issues related to these three subjects, noting the work of the leading thinkers and the themes of the seminal articles. While the legal focus is upon the United States and Canada, examples from Great Britain, the Commonwealth and Continental Europe are included where relevant. Section 1 explores the concept of decisional capacity, its historical evolution, and contemporary challenges to the established methods of assessment. Section 2 then examines the Anglo-American legal doctrine of "informed consent," focusing upon how much information a patient must be afforded when asked to render choices about medical care. Finally, Section 3 surveys medical decision-making and consent in situations in which a patient lacks the capacity to make her own choices including the use of advance directives, default agents in the absence of advance directives, and the distinct challenges raised by psychiatric, pregnant, and unrepresented patients. I am grateful to the late Edward N. Beiser of Brown University for first introducing me to these topics and to my students and colleagues at the Icahn School of Medicine in New York City and the Alden March Bioethics Institute of Albany

Medical College for regularly challenging my thinking on these issues. Needless to say, any errors or omissions are entirely my own.

2 Capacity

The concept of decisional capacity in medicine is of relatively recent origin, dating back to significant changes in both clinical ethics and in society at large that took place in the 1960s and 1970s. The evolution of the concept paralleled closely the concomitant shift from a paternalistic medical culture to one that prioritized patient autonomy. Before that time, allopathic medicine in the western world operated largely on the "assumption that physicians [were] in charge of their patients and [were] therefore entitled to make medical decisions, even for patients who wish[ed] to make their own."[1] Often, diagnoses were entirely withheld from patients, and sometimes, unbeknownst to these patients, family members joined physicians in playing a decisive role in a patient's medical care. Two major studies from the 1960s revealed the extent of these paternalistic views. In 1961, psychiatrist Donald Oken reported that nearly 90 percent of American physicians surveyed often or always kept cancer diagnoses from patients, while in 1965, sociologists Barney Glaser and Anselm Strauss found that a significant majority of doctors did not inform patients when these patients suffered from terminal illnesses.[2,3] The practice of withholding diagnoses also proved widely acceptable to patients, especially in an era in which many diseases remained highly stigmatized and in which relatively few effective medical interventions remained available. Not sharing such information was reflected in the popular culture, such as the Pollitt family's choice not to tell "Big Daddy" that he is dying of cancer in Tennessee Williams' Pulitzer Prize-winning play, *Cat on a Hot Tin Roof* (1955); in some communities, one whispered the word "cancer" with shame as one did "adultery" or "divorce."[4] Frequently, withholding diagnoses did not have any direct clinical impact upon patients themselves as the limited availability of therapeutic options meant that no medical decisions needed to be rendered at all – although one might argue that personal decisions, such as whether to retire from one's job or to write a will, might well have depended upon knowing one's impending fate in advance. The renowned British obstetrician-gynecologist, Peter Huntingford, was branded a "rebel" in 1975 for arguing, as BBC Television Science Correspondent James Wilkinson described it, that "doctors should stop behaving like gods" and "pretending that they know all the answers."[5] Yet only nine years later, Wilkinson described the era in which physicians "were often looked upon as omniscient by their patients" as ending "years ago."[6] Wilkinson's observations reflected a radical and sudden shift in medical practice. By 1979, a survey identical to Oken's (1961) study revealed that nearly all American physicians shared cancer diagnoses with their patients.[7]

Sweeping changes transformed American psychiatry during this era as well. Autonomy-friendly legal rulings in *Wyatt v. Stickney* (1972) and *O'Connor v. Donaldson* (1975) increased the power of psychiatric patients to challenge the decisions of their doctors.[8] Intellectual critiques of the supposedly controlling and authoritarian nature of psychiatric practice emerged in the writings of Canadian sociologist Erving Goffman and French historian Michel Foucault.[9] Starting in the 1950s, a generation of anti-establishment psychiatrists, including Marc Hollender and Thomas Szasz, began to call for a complete end to psychiatric paternalism and the establishment of care models based upon "mutual participation."[10,11] Initially, these efforts had only a limited impact upon the evaluation of patient decision-making. Yet as patients were increasingly offered a voice – or even a final say – in their own medical decisions, a mechanism needed to be available to determine whether these patients possessed the cognitive abilities to render appropriately the decisions in question. Drawing upon over a century of work related to the concept of legal competence (i.e., the legal ability to engage in certain transactions) in non-medical fields, a generation of physicians and philosophers, including forensic psychiatrist Loren Roth and priest-turned-bioethicist James F. Drane, crafted a set of principles and guidelines to govern such assessments. By the late-1980s, with additional contributions from psychiatrist Paul Appelbaum, psychologist Thomas Grisso, and philosophers Allen Buchanan and Dan W. Brock, among others, the concept of decisional capacity assessment had become firmly ingrained in allopathic medical practice as a key component of care in hospitals across the western world.

2.1 Classification of Capacity

The capacity to render medical decisions, often referred to as "decisional capacity," is one of many forms of capacity and among the most recent to develop. Other types of capacity assessment arose much earlier and came with their own sets of rules including testamentary capacity (i.e., the capacity to write a will), testimonial capacity (i.e., the capacity to testify in court), contractual capacity (i.e., the capacity to sign a binding and enforceable contract), and capacity to marry. Among these, case law and controversies surrounding testamentary capacity emerged first, in the mid-nineteenth century, as "legal challenges to wills became common."[12] Ryan Hall et al. attributed these challenges to three underlying causes: increased distribution of wealth, novel "medical and legal theories regarding insanity and mental illness," and "legal ambiguity regarding what constituted a sound mind."[13] Around this time, psychiatrist Isaac Ray outlined criteria for testamentary capacity including that "the memory must be active enough to bring up to mind all of those who have natural claims upon the bounty of the testator" and "to make him aware of the nature of his property"; in addition, if the testator made "bequests in certain sums," he should have known their value.[14] This approach, to a greater or lesser degree, was adopted broadly. Although some early legal decisions found testators incompetent based upon *any* evidence of insanity, most nineteenth century courts decided each case "upon its own merits," which in turn led to the incorporation of medical experts in the assessment process.[15] Analogous rubrics were developed relating to courtroom testimony, contract formation and marriage.[16] Meanwhile, the concept of competence to stand trial or to be convicted of a crime, and later to be eligible for capital punishment, became subject to its own set of legal analyses and rules.[17–18] Although often relying heavily upon medical testimony, ultimate determinations in these areas generally fell to judges and, more rarely, juries.

Until the 1970s, determinations related to the ability of patients to render medical decisions, when they did arise, were also left to the legal system. Findings of "competence" were generally global in that an individual who was unable to render decisions was declared "incompetent" for *all* purposes; these incompetent persons were often assigned guardians or conservators who had complete legal authority over *any* life choices of their wards. Competence in this sense refers to a "global" determination of abilities and related rights; such determinations can only be reversed through the courts, which may prove a challenging process. In contrast, decisional capacity is generally issue-specific and determined by physicians. (Whether one or more physician is required, and whether that physician or physicians must have specialized training in a field like psychiatry or neurology, varies considerably among jurisdictions.[19]) Reversing such determinations, when facts or circumstances offer justification, is comparatively easy, as these findings remain entirely within the hands of medical providers. Of note, existing terminology can lead to confusion, especially as the terms "capacity" and "competence" were often employed interchangeably in the medical literature until the last few decades. At present, with regard to medical decision-making, only the term "capacity" should be used to refer to decision-specific assessments by physicians, while the term "competence" should be reserved for the judgments of courts. Needless to say, a physician in the clinical setting should never call upon a consulting psychiatrist for an assessment of "competence," but only of "capacity." Finally, it is also worth noting that most consult-liaison psychiatrists in the hospital setting will only render assessments of "decisional capacity" related to healthcare.

Patients or family members seeking other forms of assessment, such as a determination of testamentary capacity to write a will, generally must hire their own forensic psychiatrists, as hospital-based psychiatric consultants are rarely forensically qualified and usually do not have training in capacity evaluation beyond clinical matters.

2.2 Rationality-Based Models of Assessment

2.2.1 Roth's "Tests of Competency"

The first serious attempts to establish a rubric for decisional capacity assessment occurred in the late 1970s. These efforts were generally motivated by a desire to protect the autonomy of patients, but may also have reflected a wish to enhance the power of physicians vis-à-vis the courts. Among the early tests proposed were Harry and Grace Olin's requirement that patients display "accurate knowledge" surrounding their condition and treatment, and Howard Owens' emphasis on patients' underlying abilities such that "a minimum of intact cognitive functions must be in evidence, including perception, comprehension, reality testing, and a sense of the reality of the self and the world."[20,21] Yet the most significant early development occurred in 1977 with publication of two companion articles in the *American Journal of Psychiatry*, "Tests of Competency to Consent to Treatment" and "Toward a Model of the Legal Doctrine of Informed Consent," by the trio of forensic psychiatrist Loren H. Roth, attorney Alan Meisel and sociologist Charles W. Lidz.[22,23] These two pieces examined existing approaches to decisional capacity, reviewing the benefits and shortcomings of each method. The authors' focus was on decision-making related to psychiatric treatment, but heavily influenced subsequent approaches to general medical care as well. According to Roth et al., "it has been our experience that competency is presumed as long as the patient modulates his or her behavior, talks in a comprehensible way, remembers what he or she is told, dresses and acts so as to appear to be in meaningful communication with the environment, and has not been declared legally incompetent."[24] While acknowledging "that there is no magical definition of competency" and that the "search for a single test of competency is a search for a Holy Grail," they nonetheless established certain fundamental requirements for any assessment tool for decisional capacity. These included: 1) that the test "can be reliably applied"; 2) that the test "is mutually acceptable or at least comprehensible to physicians, lawyers, and judges"; and 3) that the test is set at a level capable of striking an "acceptable balance between preserving individual autonomy and providing needed medical care."[25] Their survey of existing approaches in case law and clinical practice elicited five common standards. These included requiring that patients 1) express a choice; 2) that the choice evidenced will result in a "reasonable" outcome; 3) that the choice evidenced is based upon "rational" reasons; 4) that the patient is able to "understand the risks, benefits, and alternatives to treatment"; and 5) that the patient has "actual understanding" of these risks and benefits, not merely the ability to understand them.[26] Roth et al., also made the point that understanding the risks and benefits of treatment generally includes understanding the risks and benefits of "no treatment" at all. Rather than offer a normative argument that advanced any one of these approaches, Roth et al., presented a review of existing approaches and noted that such approaches were often combined in practice, with "circumstances" determining "which elements of which tests are stressed and which are underplayed."[27] Subsequent commentators, however, used this review to propose operational criteria – with Roth even joining psychiatrist Paul Appelbaum directly in developing a model, drawn heavily from this earlier framework, for consent in the research setting.[28]

2.2.2 Sliding Scale Models

Nearly a decade after Roth et al.'s survey of the medico-legal landscape, a former Catholic priest turned academic ethicist proposed a "sliding scale" for decisional capacity.[29] For James F. Drane, the "reasonableness" of a decision *in context* was essential for a capacity determination.[30] His model proposed three general categories of medical situations," and, "as the consequences flowing from patient decisions become more serious, competency standards for valid consent or refusal of consent [became] more stringent."[31] Drane's first or "least stringent" category applied to "medical decisions that are not dangerous and are objectively in the patient's best interest"; his second category, which requires more stringency, included situations where "the diagnosis is doubtful, or the condition chronic" or "if the diagnosis is certain but the treatment is more dangerous or not quite so effective." While great leeway and an assumption of capacity might be permitted for cases in the first category, Drane argued that cases in the second category demanded that the patient "be able to understand the risks and outcomes of the different options and then be able to make a decision based on this understanding."[32] Finally, the "most stringent" criteria for capacity was to be applied in situations where the patient's choices appeared "very dangerous" and "counter to both professional and public rationality."[33] In addition to influencing the subsequent discourse in the English-speaking world, Drane's approach played a dominant role in shaping models of capacity assessment subsequently adopted in Spain and Latin America.[34] His approach was not without its critics. Most notably, Charles

M. Culver and Bernard Gert have argued that Drane's model "conflated competence and rationality" and that "competence to make medical decisions is neither a necessary nor a sufficient condition for determining when it is morally justified to overrule patient's treatment decisions."[35,36] In practice, the model, which relies heavily upon the concept of reasonability, is rarely used on its own at present, but Drane's approach continues to be employed *in conjunction with* other models, such as the "four skills" model (discussed below) with specific regard to the degree of scrutiny that evaluators use when assessing particular decisions. For instance, although the "four skills" model requires some degree of assessment in all cases, whether the patient is consenting to a routine blood draw or refusing a life-saving appendectomy, in most clinical settings, the patient's choice will be examined much more closely under the latter circumstances. In this regard, Mark R. Wicclair has trenchantly offered the observation that a "stronger reason for making sure that a patient is decisionally capable should not be confused with a stronger standard of decision-making capacity."[37]

Philosophers Allen Buchanan and Dan W. Brock also advocated for a variation on the sliding scale model, which they termed "decision-relative" assessment.[38] They contrasted this method with approaches that used a "fixed minimal capacity" or threshold approach to cognition and understanding. According to Buchanan and Brock, "the standard of competence ought to vary with the expected harms or benefits to the patient of acting in accordance with a choice."[39] Among the novel arguments they advanced was that "just because a patient is competent to consent to a treatment, it does not follow that the patient is competent to refuse it, and vice versa."[40] For instance, agreeing to a low-risk, life-saving appendectomy under their system requires far less understanding than choosing to reject one. They proposed a continuum of cases based upon the nature and context of the decision that ranged from "low/minimal to high/maximal."[41] The concept of decision-specific assessment in the medical setting has now been nearly universally adopted in the English-speaking world.

2.2.3 The President's Commission

Another effort to provide a standard for capacity assessment emerged from the work of the President's Commission for the Study of Ethical Problems in Medicine and Biomedical and Behavioral Research.[42] This Commission, established by the United States Congress in 1978, consisted of twelve leading ethicists and worked from 1978 to 1983. In 1982, it published *Making Health Care Decisions*, a comprehensive volume that addressed the issue of capacity assessment extensively. The Commission adopted a decision-specific approach, noting that many patients "are capable of making some decisions but not others," and rejected an outcome-based approach, stating that "the presence or absence of capacity does not depend on a person's status or on the decision reached, but on that individual's actual functioning in situations in which a decision about health care is to be made."[43] The three-prong test that the Commission ultimately proposed required that, to be capable of rendering a decision, a patient possess "a set of values and goals," have "the ability to communicate and to understand information," and display "the ability to reason and to deliberate about one's choices."[44] Yet unlike the "values-based" approaches to assessment (discussed below), the Commission held out the possibility that, "[a]t certain outer limits, an individual's goals may be so idiosyncratic that they give rise to questions about the person's capacity for decision-making."[45] In other words, the Commission – like Drane, although maybe not to the same degree – considered the possibility of limitations upon autonomy based upon subjective notions of reasonableness.

2.2.4 The "Four Skills" Model

The most commonly used system of capacity assessment in American hospitals today was initially proposed by Paul Appelbaum and Thomas Grisso in a seminal article, "Assessing Patients' Capacities to Consent to Treatment," published in the *New England Journal of Medicine* in 1988. This article proved significant in two distinct ways. First, drawing upon existing commentary and case law in the field, the authors offered several general principles about capacity that are now accepted nearly universally in the United States and Canada. These included: 1) that "[c]ompetent patients have the right to decide whether to accept or reject proposed medical care"; 2) that in the absence of "substantive questions" about a patient's competence, physicians "should accept the patient's wishes"; 3) that capacity "may fluctuate with changes in a patient's underlying mental disorder or with unrelated factors such as fatigue, the effect of medications, or the occurrence of an unpleasant event immediately before the evaluation," so multiple efforts should be made, whenever possible, to determine capacity, and evaluation should be an ongoing process; and 4) that most determinations of decision-making capacity can be conducted by physicians without referral to the courts.[46] In addition, Appelbaum and Grisso made two additional observations that often receive short shift in clinical practice at present, namely, 1) that patients "may respond better to examiners with similar ethnic or cultural backgrounds" and 2) that following a determination that a patient lacks capacity to render a decision, efforts must be made to determine whether that capacity can be restored before the patient is deprived of autonomy.[47]

The second contribution of Appelbaum's and Grisso's article was to offer a "four skill" rubric for capacity assessment in the clinical setting. Deficiencies in *any* of these four abilities might be sufficient for a finding of incapacity. First, patients needed to "communicate stable choices" regarding their care that endured "long enough for them to be implemented."[48] Such a requirement did not mean that patients could not change their minds; in fact, reconsidering a decision is absolutely appropriate as circumstances evolve or as a patient's understanding of these circumstances develops. At the same time, repeat and rapids reversals of course may be an indication of either psychosis or an affective disorder.[49] In the extreme, such vacillation is characteristic of aboulomania, a pathological condition in which patients suffer from debilitating indecisiveness.[50] Second, the patient needed to "understand information about a treatment decision" including "risks and benefits" and "likelihood of various outcomes."[51] Third, the patient had to appreciate this information as it related to them personally, including "acknowledging illness when it is shown to be present, evaluating its effect and the effect of the treatment options presented, and acknowledging that the general probabilities of risks and benefits apply to the situation."[52] Clinicians must be careful to ascertain that the patient recognizes the personal applicability of the information: For instance, a patient may express the belief that antibiotics can cure bacterial pneumonia, but unless asked directly, may not also reveal that he believes these antibiotics unnecessary in his own case, because he is protected by a magical forcefield. Finally, the fourth skill was that the patient needed to "manipulate information rationally," a catchall concept that included the ability to weigh "multiple options simultaneously, in a way that reflects the weights previously assigned to them" and to reach logical conclusions derived from starting premises.[53] Under the Appelbaum and Grisso model, patients able to demonstrate these four skills retained their decisional capacity.

2.2.5 Temporality

Since decisional-capacity may evolve with time, assessments should be conducted proximally in time to the proposed intervention. No hard and fast rule exists regarding how much time may elapse before a reassessment is required; instead, a rule of reasonableness seems appropriate – with the time-frame being reduced as the stakes of the decision increase or if there is reason to believe that the patient's views have altered as the result of an intervening event. However, a strong consensus exists – recognized in some statues[54] – that physicians cannot wait for a patient with capacity to deteriorate in order to overrule a clear decision. For example, if a patient presents to the hospital declining emergency dialysis, and possesses capacity to render that choice, a nephrologist may not ethically defer care until such a patient develops uremia (a build-up of nitrogenous waste-products in the blood that leads to confusion) in order to then override the prior decision on the grounds that patient no longer possesses capacity. Such an approach is too clever by half and is likely to result in civil liability.

2.3 Challenges to Existing Models

A range of challenges to the dominant "four skills" approach to capacity have emerged in the twenty-first century as decades of practice have led to increased empirical evidence regarding how this model impacts patients in clinical settings. Three leading critiques are examined below.

2.3.1 Tool v. Intervention

Since its development in the 1970s and 1980s, capacity assessment has generally been thought of as a neutral tool akin to other minimal-risk diagnostic tests like Mary H. S. Hayes and Donald G. Patterson's Visual Analogue Scale for rating pain or Hermann Snellen's use of an eponymous chart to assess visual acuity. Many providers continue to view such assessments as entirely benign. However, Jacob M. Appel and Omar Mirza have posited that capacity assessments are clinical interventions with appreciable risks and potentially negative consequences of their own – more akin to invasive diagnostic procedures.[55] They describe such assessments as "capacity challenges" and believe them to be "high stakes" encounters which place "[i]n jeopardy . . . a patient's autonomy and the right to render her or his own medical decisions."[56] These assessments "are often conducted by psychiatric consultants who are meeting the patients for the first time."[57] Far from harmless, encounters with psychiatrists can be "stigmatizing," a risk that is "elevated among historically marginalized populations whose communities have been victimized through systematic mistreatment by mental health professionals."[58] Appel and Mirza note that "patients undergoing capacity challenges are often in a highly vulnerable state" and that a capacity assessment may "exacerbate a patient's sense of powerlessness."[59] In addition, capacity challenges "risk furthering mistrust and undermining the therapeutic relationship with the care team," which in turn "may make the patient less likely to consent to the underlying proposed medical intervention."[60] Alternatively, the patient may consent at present, but at a future encounter with the healthcare system, the patient may refuse to engage with a capacity assessment and find her autonomy curtailed as a result. Of course, all medical interventions run *some* risk of damaging the therapeutic relationship if they go poorly. Yet one feature that distinguishes capacity challenge from nearly all other physician–patient interactions in medicine is that no effort is rendered, either formally or informally, to provide informed consent to the patient regarding the capacity challenge itself – nor have any such efforts been documented in the literature. In short, capacity challenges may have more significant implications, both clinically and psychologically, then providers recognize.

That does not necessarily mean that physicians should never inquire into a patient's decisional capacity. In our current healthcare delivery framework, such assessment is often essential. However, physicians have an obligation to minimize risks – just as they do with any other medical intervention. First, every effort should be made to conduct capacity assessments or request capacity consults only when doing so is of clear clinical value. For example, if the results of such a challenge will have no bearing upon the future course of treatment – either because the physicians will not perform the underlying intervention without both patient consent *and* cooperation, or because a third-party decision-maker, such as a proxy or surrogate, is in full agreement with the patient's choice and will implement her wishes even if she is determined to lack capacity (based upon her known prior preferences) – then such a superfluous assessment should not occur. Second, care should be taken to reduce risks when a capacity challenge is necessary. Among steps that might mitigate such risk are carefully explaining the purpose of the assessment and, when possible, having someone with whom the patient is comfortable conduct the evaluation. Psychiatric engagement should be a last resort, especially in patients already suspicious of mental health professionals. If a psychiatrist must be consulted, the physician caring for the patient should explain to the patient in advance why a psychiatrist has been called upon and the nature of the evaluation process.

2.3.2 The Primacy of Rationality

The most frequent critique of the "four skills" model and similar approaches to capacity assessment emphasizes its overreliance upon rationality. For many patients, to apply the reasoning of the late New York State Justice Sol Wachtler, "deciding the issue is substantially easier than explaining it."[61] Yet the "four skills" model denies autonomy to these very patients if they are ultimately unable to offer a logical justification for their preferences. As Derek Morgan and Kenenth Veitch have observed, the assessment is less about the ability to render a choice and more about "whether the person making that decision can construct a convincing case why he or she reaches the standard of the 'ability' that law expects in such circumstances."[62] In fact, Eilionóir Flynn makes the unsettling observation that, according to psychological data, "individuals only conjure up reasons for their decisions when called upon to do so, and these reasons rarely correlate with their actual decision-making process at the time of the original decision, but rather reflect the most persuasive explanation the person can find for his/her decision."[63] Yet common sense suggests that knowing what one wants and being able to explain the underlying reasons for one's preferences are two fundamentally different matters. Appelbaum and Grisso, according to critics, fail to offer a convincing case for why the latter should receive primacy over the former.

2.3.3 Social Capital and Bias

The bioethics principle of justice requires that people in similar situations be treated similarly. An ideal capacity assessment process would ensure consistency between evaluators (interrater reliability) as well as an accurate and unbiased assessment of each patient (internal validity). Unfortunately, multiple studies have shown high rates of inconsistency between capacity evaluators when the "four skills" model is used.[64–65] In addition, increasing evidence suggests that the consequences of this variability do not impact all populations equally. Deep, systemic bias against minority and low-income populations has long been a feature of the healthcare system in the United States.[66] These biases played a formative role in the development of American psychiatry and, not surprisingly, continue to influence current clinical practices.[67,68] For example, overwhelming data supports the presence of racial bias in the inaccurate over-diagnosis of psychotic disorders in Black populations.[69] As one might expect, such biases plague the process of capacity assessment as well. A pioneering study by William Garrett et al. found that at a major, tertiary care center in New York City, "significant racial disparities" occurred "within decisional capacity consultation placements" in that far more unactionable (and hence unnecessary) consultation requests were called for Black and Hispanic patients than for white and Asian patients.[70] If capacity assessment is thought of as an intervention rather than a tool, then these unactionable evaluations imposed a serious and unjustified burden upon Black and Hispanic populations. Whether similar biases in capacity exist with regard to economic status, gender, or other factors related to social capital and vulnerability remains a question for future investigation. What can be said with confidence is that, at a time when the medical and mental health professions are struggling to overcome legacies of structural bias, the existing use of dominant capacity assessment tools appears to be exacerbating those biases.

2.4 Values-Based Models of Assessment

The "four skills" model of capacity assessment was principally designed for patients who likely met the standards for decisional capacity at a recent point in the past but now appeared to have deviated from that baseline: Such patients might include those suffering from intoxication, delirium, psychosis, and dementia. In contrast, the model was not designed for individuals whose long-term personal, cultural, or religious beliefs or preferences differed so significantly at baseline from those of allopathic physicians that they likely had never met the standards for decisional capacity established by the "four skills" model. For example, adherents to the Church of Christ, Scientist (i.e., "Christian Scientists"), a religion founded in 1879 that now claims 400,000 members worldwide, "rely exclusively on faith healing as a response to infirmity" and reject the efficacy (as opposed to the morality) of modern medical treatment.[71] Nevertheless, adult Christian Scientists are generally permitted to make their own medical decisions in North American and European hospitals, requiring an exception to the "four skills" model.

The "four skills" model also accounts poorly for several common coping mechanisms, such as "denial," which can prove healthy for patients under some circumstances.[72] The elderly patient with a low prognosis of overcoming an aggressive metastatic cancer may insist that her pain is merely a result of rheumatism and refuse to engage with questions about chemotherapy, even if such care offers a small chance of extending survival. Many providers, however, would prove reluctant to overrule a patient's refusal of chemotherapy under such circumstances based upon a failure to meet the "four skills" standards. As a result, such cases are often "punted": decided upon the artificial premise that the cost-to-benefit balance of forcing such care upon the patient would lead to unjustifiable distress. Once again, the limitations of the "four skills" model become apparent. To make the model both fully functional and consistent with justice in a clinical setting, the number of exceptions required proves extensive – reminiscent of the baroque scheme of extra orbits and epicycles that Ptolemy required to model a geocentric solar system.

In response to shortcomings of the "four skills" model, American psychiatrists Jacob M. Appel and Omar Mirza have proposed an alternative, "patient-centered" approach to capacity assessment designed to reduce the number of unnecessary capacity evaluations and to minimize the need for exceptions. Appel argues that while in theory, "patients are assumed to have autonomy to render their own medical decisions until proven otherwise," the current application of the four skills model compels "the patient [to] defend her decision with reasoned argument" and penalizes her if she cannot do so.[73] Like the "four skills" model, the "values-based model" involves four steps: First, a clinician must ascertain the patient's "underlying and longstanding values" including "attitudes toward medical care and, as applicable in the patient's circumstances, toward other related matters that may have an impact upon care: religion, spirituality, family, quality of life, death, pain, etc."[74] These values can be assessed through direct dialogue with the patient, through the insights of friends and family members, and through evidence previously documented in either advance directives or the medical record. Second, the clinician must compare these underlying values with the patient's preferences as currently expressed. If the two are concordant, the assessment of decisional capacity should conclude, and the patient be permitted to render her own decision. When the patient's current preferences prove consistent with underlying values, any basis for this consistency is not relevant – even if the patient arrives at the "right" answer for an illogical reason. Third, if the underlying values and current preferences diverge, the clinician must explore the reasons for this non-concordance, looking for intervening events, such as a religious conversion or a "deliberate and systematic recalibration of personal values."[75] If the explanation for the divergence is "coherent and plausible," the assessment of decisional capacity should again conclude, and the patient be permitted to render her own decision. Finally, if the patient's underlying values and current preferences are not aligned, but in a manner that cannot be explained coherently, then the prospect of restoring the patient's cognitive or psychiatric condition to the point where either past values and present preferences align, or the divergence can be explained, should be assessed. If restorability appears possible, it ought to be attempted. Only if restorability is not likely should "the patient should be deemed to lack capacity and the appropriate third-party decision-maker should be consulted."[76] Values-based assessment arguably addresses many, but not all, of the biases in the "four skills" model. However, it has not yet been widely adopted in practice.

Even should clinicians seek to replace the "four skills" model with a "values-based" model, significant barriers exist to doing so. Some of these are cultural: Many practitioners may resist shifting away from a widely accepted standard, barring highly persuasive evidence that it is less effective, which is hard to demonstrate in a relatively subjective area like capacity assessment. Inertia is often a powerful systemic force in these regards. Another barrier that would have to be surmounted is the existence of statutes in many jurisdictions that have codified all or part of the "four skills" model into law. Until the 1970s, no

American state legislature had attempted to define decisional capacity. Idaho, which did so in 1977, adopted a very broad "comprehensibility standard."[77] Yet over the succeeding four decades, the vast majority of American state legislatures have ventured into the field. The National Conference of Commissioners on Uniform State Laws proposed a Uniform Health-Care Decisions Act (UHCDA) in 1993, which included a definition of decisional capacity that, while only adopted verbatim by six states, did spur many other jurisdictions into action. The UHCDA closely followed the "four skills" model, defining capacity as "an individual's ability to understand the significant benefits, risks, and alternatives to proposed health care and to make and communicate a healthcare decision."[78]

At present, forty-two US states and the District of Columbia "define decisional capacity by statute," while only "nine American jurisdictions do not."[79] Of the forty-two jurisdictions that have codified a definition of capacity, ten include three or four of the "four skills," twenty-three include two, and another six include one.[80] (Of note, several of these states have definitions that apply only if the patient has executed an advance directive.) Most important, thirty-two of those forty-two jurisdictions with statutes have adopted definitions narrow enough that overt legislative action would be required to shift to a "values-based" approach.[81] These same obstacles are likely to arise outside the United States as well. Ontario, Canada, for example, adopted a Health Care Consent Act in 1996 that requires patients to "understand" and "appreciate" decisions in order to establish capacity.[82] In Great Britain, the Mental Capacity Act, effective as of 2007, imposed a four-skills approach very similar to that proposed by Appelbaum and Grisso: The patient must "communicate the decision," "understand relevant information," "retain that information" and "use or weigh the information as part of the process of decision making," How amenable lawmakers will be to adjusting these definitions remains an open question.[83]

2.5 Special Cases and Novel Challenges

2.5.1 Volitional Non-communicators

One particular challenge that arises frequently in the assessment of decisional capacity involves patients who volitionally opt against expressing any choice. The "four skills" model was not designed with such cases in mind. Appelbaum and Grisso speak of the "ability to communicate choices," not the willingness to do so.[84] Nevertheless, their "model that has become so ingrained in medical practice that clinicians and ethicists adopt this approach reflexively in cases for which it was not designed" on a regular basis.[85] As Samia Hurst has observed, current practice is often to accept that "it cannot be concluded that such patients are incompetent," but still "treat them as if they were."[86] Patients may wish to withhold their choices for a range of reasons including holding a "philosophical, religious, or cultural objection to acknowledging the role or authority of the medical team," an unwillingness "to concede the legitimacy of allopathic medicine," a "lack confidence or trust in the healthcare system or their doctors," "[a]nger or hostility toward the care team," or a "personality pathology . . . [or] history of trauma that renders engagement emotionally difficult."[87] Finally, some patients may have private reasons for refusing to engage, both valid and rational, that they are simply unwilling to divulge. Patients who have had negative personal experiences with the healthcare system and those who have endured the consequences of systemic bias may be particularly unwilling to engage.[88]

While the "four skills" model is clearly not appropriate for addressing such situations, value based models will also prove deficient, so clinicians face the dilemma of how to proceed. One recent proposal suggests a seven-step process for approaching such patients with the ultimate goal of persuading them to engage.[89] These steps include: 1) Giving the patient maximum time to reconsider and ensuing that "efforts to evaluate the patient . . . occur at different times and on different days" to be confident that "a fleeting emotion is not preventing the patient from engaging"; 2) employing multiple different evaluators, including non-clinicians such as chaplains and social workers, and striving for demographic concordance between evaluator and patient; 3) engaging third parties, such as relatives, in the process to build trust; 4) affording protection from "emotional injury or distress," which may even "require an apology by hospital staff designed to placate the patient's frustrations – whether justified or not – and to allow him to save face and avoid narcissistic injury"; 5) clarifying the consequences of not expressing a choice; 6) exploring "the circumstances of the refusal to engage, as well as his statements and behavior, to determine whether it gives strong indication of the patient's wishes"; and 7) as a default, when all efforts to engage prove fruitless, assuming "that refusal to engage means refusal to accept intervention," rather than concluding that the patient lacks capacity.[90]

2.5.2 Neuroimaging and Capacity

A traditional understanding of capacity held that an individual diagnosed as being in a persistent vegetative state (PVS) lacked the ability to express himself and therefore, by definition, also lacked the legal authority to render his own medical decisions. Recent research by British neuroscientist Adrian Owen has challenged that understanding. Using fMRI technology, Owen has shown that some PVS patients not only possess awareness, but are actually able to communicate answers to yes or no questions. As Owen described the results of his study, after asking a patient to imagine "playing tennis to convey a 'yes' response" and to imagine "walking around his house to convey a 'no' response, this patient was able to communicate the answers to a series of biographical questions, such as whether he had brothers or sisters and the last place he'd visited before his accident 5 years earlier – all pieces of information that we did not know at the time but could verify as being correct with the family at a later date."[91] In the short term, the implications of this research have the potential to transform capacity assessment for a subset of patients. If such patients can answer biographical questions, presumably they can also provide yes or no answers to questions about medical decision-making – even, in theory, whether to continue life support. At the same time, to what extent such limited communications can ensure the evaluator that the patient has a full understanding of his decisions remains uncertain.

Far more significant are the potential implications of neuroimaging for capacity more generally. As technology advances, yet unknown methods may be developed to harness neuroimaging techniques to tap into the underlying values or preferences of patients that cannot be extracted with traditional methods of assessment. Rather than a static set of assessment tools, capacity evaluation remains a dynamic field in which both clinical technology and ethical understanding continue to evolve.

3 Informed Consent

The notion that patients must agree to interventions performed upon them by physicians likely dates back to ancient times, but efforts to impose this obligation through legal mechanisms did not emerge until the late nineteenth and early twentieth centuries. That is not necessarily to say that in an earlier era, physicians never deferred to the wishes of their patients. In fact, considerable debate exists in the literature regarding the prevalence of such practices. Medical historians including Martin Pernik and Kathy Powderly have argued that such deference and respect for patient autonomy may have been widespread.[92] Pioneering yellow fever researcher Walter Reed used informed consent forms for human subjects in Cuba as early as 1900.[93] Nevertheless, the courts did not yet recognize such safeguards as a requirement of practice. In an earlier era, the criminal law had imposed at least a limited check upon rogue medical men who forced surgery onto unwilling citizens though statutes banning battery and assault. But starting at the end of the Victorian Age, courts began to enforce such limits on medical authority through civil litigation. Later, consent to care alone proved insufficient to avoid liability. Rather, courts began to impose "an explicit duty to disclose certain forms of information and to obtain consent in both practice and research."[94] The term "informed consent" itself, when used in this context, dates from the 1950s.[95] Since that time, it has come to be used in similar ways in both clinical and investigative settings, although this section will confine itself to the uses of "informed consent" as it relates to consent to treatment. By the 1970s, the doctrine of "informed consent" had become firmly embedded in American healthcare and the failure to obtain meaningful informed consent widely recognized as a basis for malpractice payouts.[96] The American Medical Association's Code of Ethics now states: "The patient's right of self-decision can be effectively exercised only if the patient possesses enough information to enable an informed choice" and that a "physician's obligation is to present the medical facts accurately to the patient or to the individual responsible for the patient's care and to make recommendations for management in accordance with good medical practice."[97]

Definitions

Defining informed consent in the clinical arena has posed an ongoing challenge for both courts and ethicists. In a previous section on capacity, this volume examined the level of understanding required to agree to care. A related question is how much information such an individual must receive prior to rendering a decision. For instance, in sharing potential risks, which risks must be shared with the patient? Clearly, not sharing *any* of the known risks – even those that are highly likely – will not pass legal or ethical muster. Yet some risks are so far-fetched that no clinician might reasonably be expected to disclose them. For instance, explaining to a prospective surgical patient that a comet might hit the hospital while he is under anesthesia, disrupting an operation and resulting in a poor outcome, approaches absurdity, even though collision with such a cosmic object is certainly a theoretical possibility. Between these two extremes, courts have struggled to determine what level of information sharing should suffice to meet the medical standard of care. However, the quantity and quality of information shared is only a portion of the equation. The law must also decide the degree, if any, to which the patient must *actually* understand that information. In other words, is the standard an objective one: namely, that the physician related the information that a physician should reasonably be expected to share under the circumstances? Or is the standard a *subjective* one that requires the patient to actually internalize and understand this information? To the extent that courts adopt an objective standard, "informed consent" becomes a misnomer that is better understood as "reasonable disclosure." It is also worth noting that the objective standard may prove much easier for judges and juries to operationalize. Some cosmetic surgeons now even videotape their informed consent processes so empirical evidence exists regarding what information was shared with prospective clients. In contrast, a subjective standard may protect patients, especially vulnerable individuals who genuinely do not possess an understanding or appreciation of the divulged risks. However, a subjective standard poses several drawbacks beyond increased administrative costs for the legal system in enforcing it. First, some patients may perjure themselves to claim a failure to understand when they are unsatisfied with medical outcomes. Second, patients – like everyone else – may be subject to so-called "Monday morning quarterbacking" in which they reexamine their choices in light of the outcomes, and assert, with sincerity, that if they had "really" understood the risks, now that they are unhappy with the results, they would not have consented in advance. After all, hindsight is twenty-twenty. Disproving such claims is nearly impossible, which is why most jurisdictions have shifted toward an objective standard.

3.1 Historical Evolution of Informed Consent

3.1.1 Early Cases

The roots of the informed consent principle in American law can be traced at least as far back as the United States Supreme Court's decision in *Union Pacific Railway Company v. Botsford* (1891).[98] The question adjudicated in that seminal case was whether "in a civil action for an injury to the person," a court "may order the plaintiff without his or her consent, to submit to a surgical examination as to the extent of the injury sued for."[99] The basic facts of the case were not in dispute: Clara Botsford had been travelling on a Union Pacific train when a sleeping car berth collapsed and hit her on the head. What was disputed were the extent of her injuries. Writing for a 7–2 majority, Supreme Court Justice Horace Grey found that the right "to be let alone" embraced a right to basic bodily integrity, even in civil litigation initiated by the party resisting internal examination.[100] According to Grey, "No right is held more sacred, or is more carefully guarded by the common law, than the right of every individual to the possession and control of his own person, free from all restraint or interference of others, unless by clear and unquestionable authority of law." As a result of *Botsford*, the consent element of informed consent found a foothold in American civil jurisprudence.

Another decade elapsed before the American courts began to apply the general principle announced in *Botsford* to litigation against physicians. The first case to do so appears to be that of *Mohr v. Williams (1905)*, in which the Minnesota Supreme Court ruled that surgeon Cornelius Williams was liable for performing an ossiculectomy upon Anna Mohr without her express permission.[101] Mohr had presented complaining of difficulty hearing in her right ear and had consented to surgical intervention. However, while his patient was under anesthesia, Williams determined that her *left* ear was in need of surgery while her right ear was not, so changed ears without reconsenting her. Although this operation "was in every way skillful and successfully per-formed," and may even have "benefited the plaintiff by curing her disease," she believed that her hearing had been impaired by the procedure and brought suit.[102,103] Writing for the court, Justice Calvin Brown noted that the individual has "a right to complete immunity of his person from physical interference of others, except in so far as contact may be necessary under the general doctrine of privilege; and any unauthorized touching of the person of another, except it be in the spirit of pleasantry, constitutes an assault and battery."[104] The medical context did not absolve Dr. Williams of this general obligation and he had no special privilege as a physician to override it. That Willaims had operated with the best of intentions did not sway the court, nor were they willing to infer a presumption of consent for one procedure to result from consenting for another.

Shortly thereafter, the Illinois Supreme Court came to a similar conclusion in *Pratt v. Davis* (1906).[105] (In fact, the *Mohr* court even cited a lower court ruling in *Pratt* that had taken place earlier.)[106] In that case, Dr. Edwin Pratt of Chicago had performed a hysterectomy on Mrs. Parmelia J. Davis without her permis-sion in an abortive effort to cure her epilepsy. The facts of the case were not in question as Pratt acknowledged "deliberately and calmly deceiving the woman" by not telling her "the whole truth" about the surgery.[107] In her lawsuit, Davis did not contend that she had been harmed during this operation, but rather that she was owed punitive damages for the non-consented surgery, even in the absence of actual damages. A lower court judge agreed, awarding a significant payout to Davis. Justice Guy C. Scott upheld the verdict. He found that even in the absence of tangible damages, "pain and suffering" following "the removal of the uterus," can be inferred, and Davis had failed to consent in advance to such pain.[108] In non-emergency situations, Scott ruled, a physician could not change the course of treatment without approval of a competent patent. In the words of the court, "The consent of the patient should be a prerequisite to a surgical operation where he is in possession of his mental faculties and well enough to consult about his condition without dangerous consequences to his health, and where no emergency exists making it impracticable to confer with him or requiring immediate action for the preservation of life or limb."[109] Alas, in a tragic turn of events, Davis died before the appellate court handed down the decision in her favor.

The principles of *Mohr* and *Pratt* were extended even further in the Oklahoma case of *Rolater v. Strain* (1911).[110] Dr. J. B. Rolater, the operator of a sanitarium in Oklahoma City, performed an operation on a telephone company employee, Mattie Inez Strain, who had stepped upon a nail and suffered a likely infection. Strain consented to an operation in which "an incision" was to be made "in the foot or toe so as to drain the joint and remove any foreign matter that might be found therein," but was also assured by Rolater that he would not remove any of her bones.[111] However, during the procedure, Rolater determined that removal of the "sesamoid bone" was essential to a successful intervention – so he extracted the bone without Strain's permission. In response, Strain sued for battery. Citing both *Mohr* and *Pratt*, Judge Clinton Galbraith expanded upon these earlier cases in finding the plaintiff liable on the grounds that he "had not performed the procedure in the manner agreed upon between the physician and patient."[112] In other words, it was not merely the outcome regarding which the patient had to offer consent, but at least to some degree, the method was also legally relevant. As discussed further below, the extent to which a patient can guide the manner of a medical intervention remains a complex ethical and legal challenge even to the present day.

These three rulings formed the basis for the most well-known – and, arguably, most misunderstood – case in the history of informed consent law: *Schloendorff v. Society of New York Hospital* (1914).[113] For many years, the facts of the case appeared to be clear. A 56-year-old voice coach from San Francisco, known variously as Mary Schloendorff and Mary Gamble, moved to New York City after the 1906 earthquake and presented to New York Hospital in Manhattan for treatment of "stomach pain and severe weight loss, which she attributed to anxiety resulting from the earthquake."[114] She was diagnosed with uterine fibroids and subsequently transferred from the medical service to the surgical service, where she allegedly emphasized to the nursing staff that she did "she did not want an operation."[115] Nevertheless, she received a hysterectomy. As a result of complications from the procedure, she also ultimately lost the tips of her thumb and forefinger. Schloendorff subsequently sued the hospital, "claiming $50,000 in damages for the loss of her fingers and her pain and suffering," but not for the loss of her uterus.[116] Of interest, many of these widely believed facts about the case have recently come into question. For instance, Chervenak et al. noted that the "widely held belief that Ms Schloendorff did not consent was never adjudicated to be a matter of fact" and have argued convincingly that while "[i]t is assumed commonly that Ms. Schloendorff was the victim of an intentional battery because she was anesthetized already when a pelvic mass was discovered, … in truth she may have been victimized by a lack of communication among clinicians who would have legally and ethically respected her wishes not to have surgery, if indeed that had been her expressed decision."[117] The question before the New York Court of Appeals – the highest court in that state – was whether to overturn a directed verdict against Ms. Schloendorff; in other words, the court was not deciding upon the merits of her case, just its plausibility. Justice Benjamin Cardozo, writing for the court, assumed that a battery had occurred, as is the judge's role when evaluating the merits of such a directed verdict, and based upon that non-litigated premise, observed that, "Every human being of adult years and sound mind has a right to determine what shall be done with his own body; and a surgeon who performs an operation without his patient's consent, commits an assault, for which he is liable in damages."[118] Like Judge Galbraith in *Rolater*, he cited both *Mohr* and *Pratt*. More than a century later, the case is remembered for this broad principle, rather than his attendant ruling that the hospital enjoyed charitable immunity, preventing Schloendorff from recovering so much as a dime.[119] Not until *Bing v. Thunig* (1957) would New York revisit this broad exemption for hospitals from liability.[120]

3.1.2 The Modern Era

Although the principle of consent was established by mid-century, the parameters of such consent were not fully flushed out until well after World War II. The legacy of Nazi experimentation during the Second World War, widely publicized in the "Doctors' Trial" at Nuremberg in 1946 and 1947, led to the promulgation of the Nuremberg Code, which required the "voluntary consent of the human subject" and that the research subject "should have sufficient knowledge and comprehension of the elements of the subject matter involved as to enable him to make an understanding and enlightened decision."[121] For researchers, at least, violations of informed consent could prove the basis for crimes against humanity. The trial also helped shape public thinking regarding the rights of patients in clinical settings. Yet as late as the mid-1950s, some American courts did not require full disclosure of relevant information in the consent process.[122] For instance, in *Hunt v. Bradshaw* (1955), the physician "advised the plaintiff the operation was simple, whereas it was serious and involved undisclosed risks," and the patient lost his arm during the procedures, yet the Supreme Court of North Carolina found no obligation to disclose the risks of the surgery, so long as what the doctor did divulge were not "false" to his knowledge.[123] Yet starting with the California Supreme Court's decision in *Salgo v. Leland Stanford Jr. University Board of Trustees* (1957), state courts began to impose specific standards relating to consent upon physicians in the field.[124] In *Salgo*, Dr. Frank Gerbode diagnosed Martin Salgo with "a probable occlusion of the abdominal aorta which had impaired the blood supply to the legs" and admitted Salgo to the hospital to perform an angiogram. Unfortunately, Salgo became paralyzed from the waist downward as a result of the procedure. Among his claims in court, Salgo argued that Dr. Gerbode had an obligation to warn him of this potential complication in advance and that "the details of the procedure and the possible dangers therefrom were not explained."[125] Judge Absalom F. Bray agreed, writing, "A physician violates his duty . . . if he withholds any facts 'which are [necessary to form the basis of an intelligent consent by the patient to the proposed treatment" and "[l]ikewise the physician may not minimize the known dangers of a procedure or operation in order to induce his patient's consent."[126] (Whether this knowledge of the risks of the angiogram would actually have resulted in Salgo refusing the procedure went unconsidered, although in practice, most patients in Salgo's circumstances would have accepted the risks rather than the far more dangerous consequences of an untreated Aortic blockage.)

American courts adopted this principle in a series of subsequent cases including, most notably, *Mitchell v. Robinson* (1960), *Natanson v. Kline* (1960), and *Cobbs v. Grant* (1972)[127–128] In contrast, England and Scotland initially did not embrace an informed consent standard directly, rather encompassing the principle within the broader one of negligent care.[129] In Great Britain, as developed in *Bolam v. Friern Hospital Management Committee* (1957), the failure to warn of risks could only be used as a basis for negligence if *all* reasonable physicians would be expected to disclose these risks to a patient under similar circumstances.[130] In fact, the doctrine of informed consent did not fully emerge in Great Britain until the early 1980s with the ruling in *Chatterton v. Gerson and Anor* (1980).[131]

3.2 Standards of Care

The standard of care refers to the conduct required of a physician to avoid civil liability. How much information a physician must legally share in the informed consent process reflects the medical standard of care in this regard. Over the past half century, the criteria for this standard have evolved rapidly. Starting in the early 1970s, a series of cases shifted the generally accepted standard from one that was local and descriptive to one that was national and prescriptive. As a result, the requirements for informed consent shifted as well. Increasingly, courts rejected the "professional practice" standard of a paternalistic medical culture in favor of one that demanded patients receive the information that a reasonable patient would want to know.

3.2.1 Local v. National Standard

The standard of care for medical malpractice, including for informed consent, was historically based upon the conduct of other physicians. This so-called "professional practice" rule required that a physician demonstrate the same skills as other clinicians. But which physicians was the original doctor to be compared with – those in his local community or the most talented providers in the nation? The Supreme Court of Massachusetts attempted to resolve this question in the case of *Small v. Howard* (1880), which revolved around the degree of skill needed to dress a wounded wrist.[132] According to the court, a physician "in a town of comparatively small population" is obligated "to possess that skill only which physicians and surgeons of ordinary ability and skill, practicing in similar localities ... and he was not bound to possess that high degree of art and skill possessed by eminent surgeons practicing in large cities."[133] The purpose of the locality rule was to protect rural physicians "in a time when rural and urban physicians may have had vastly different experiences with respect to their education, training, and ability to obtain the latest information relating to diagnosis and treatment."[134] One consequence was to exclude the testimony of expert witness from outside the local community.[135] Such an exclusion often made obtaining *any* expert impossible for plaintiffs, as local physicians proved loathe to testify against each other. By the mid-twentieth century, accrediting bodies had imposed national standards on most physicians, and the differences between urban and rural care had declined accordingly, turning the locality rule into an anachronism that often shielded physicians providing deficient care. As a result, courts began to impose either a statewide or national standard upon all clinicians. Massachusetts, the birthplace of the locality rule, also became one of the first states to reject it in *Brune v. Belinkoff* (1968).[136] The change occurred rapidly: As recently as 2007, twenty-one American states still followed the locality rule, while by 2014, all but five had adopted a national standard.[137] The locality principle was never embraced in Great Britain or most other common law nations, which have always required physicians to practice at the level of their peers at a national level.[138]

A national standard of care requires that physicians possess skills and knowledge consummate with their colleagues, but obviously not the same resources. A small clinic in Nome, Alaska, cannot be expected to have access to the same advanced technologies as a university-affiliated hospital in

Anchorage or San Francisco. Rather, to the degree possible, patients in locations with limited access to high-end technologies must be referred to centers that do possess them. Unfortunately, making use of such referrals often requires resources, raising equity considerations for patients endowed with fewer funds or limited social capital.

3.2.2 Prescription vs. Description

Another question that courts historically faced in medical malpractice lawsuits that bears on the standards for informed consent is whether the standard of care is merely a reflection of what physicians do in practice, or whether what they *should* do – a normative element – is also relevant. Until the 1970s, such cases were largely governed by the "professional practice" principle; as long as a physician could demonstrate that his conduct was consistent with those of other practitioners, he did not have to fear liability. As a result, establishing a violation of this standard required expert witnesses to testify that the doctor's conduct did not comport with those of his colleagues, itself a challenging barrier for potential plaintiffs in an age when physicians banded together to protect each other's interests. This "white wall of silence" operated in much the way that a supposed "blue wall of silence" has historically kept police officers from testifying against one another. Yet tort law more generally had advanced past a descriptive standard toward a prescriptive one many years earlier. As Justice Learned Hand wrote in the case of the in 1932, "a whole calling may have unduly lagged in the adoption of new and available devices," so "courts must in the end say what is required; there are precautions so imperative that even their universal disregard will not excuse their omission."[139] Yet four decades later, the courts had proven reluctant to apply this approach to the realm of medicine.

Relying on the professional practice standard was first seriously called into question in the case of *Helling v. Carey* (1974), which has been described as "the most infamous of all malpractice cases."[140,141] In this Washington state case, Dr. Thomas F. Carey had failed to detect primary open angle glaucoma, a treatable condition, in patient Barbara Helling, for a period of nine years, allegedly resulting in "severe and permanent damage to her eyes."[142] Although a routine pressure test for the condition was readily available, low-cost and easy to administer, no providers in the area of Dr. Carey's practice – at a time when the locality rule remained in effect – offered such tests to patients under age forty because glaucoma was a rare disease in this population. To the majority of the court, relying upon this industry standard was insufficient. Justice Robert Hunter wrote that, notwithstanding the general practice of ophthalmologists in

Washington State, "the reasonable standard that should have been followed ... was the timely giving of this simple, harmless pressure test."[143] Many states have since adopted this prescriptive approach or have augmented a descriptive, professional practice standard with a prescriptive rule in extreme circumstances. The result in *Helling* set the groundwork for establishing a similar prescriptive standard regarding the appropriate amount of information to share with a patient. No longer, if the general practice was to withhold information, could a physician rely upon that custom in choosing not to disclose. Informed consent, like all other areas of malpractice, developed a prescriptive as well as a descriptive element.

3.2.3 The Rejection of Professional Practice

The rejection of a "professional practice" standard in general medical practice soon led to a similar rejection in matters of informed consent. A decision of the United States Court of Appeals for the District of Columbia Circuit, *Canterbury vs. Spence* (1972), proved transformative in this regard.[144] In that case, a 19-year-old FBI file clerk, Jerry Canterbury, presented to Washington, DC neurosurgeon William T. Spence, for back pain and was determined to require a routine laminectomy. The initial procedure went well, but Canterbury fell from his bed during post-operative care, possibly as a result of the absence of a "side rail to prevent the fall," and ended up partially paralyzed. Nearly a decade after the operation, Canterbury "required crutches to walk, still suffered from urinary incontinence and paralysis of the bowels, and wore a penile clamp."[145] Whether the operation or the fall caused these disabilities was never fully ascertained. Among the many issues for the appellate court to determine was whether Dr. Spence had a duty to warn Canterbury of the risk of paralysis, estimated to be approximately 1 percent. In ruling in favor of Canterbury, Judge Spottswood William Robinson III – arguably the most plaintiff-friendly judge ever to serve on this tribunal – rejected a "professional practice" standard that would have required Spence to divulge only what other surgeons did under similar circumstances. Rather, according to Robinson, the physician was obliged to make a "reasonable disclosure" of the information that an "average, reasonable patient" would "consider important to his decision."[146] As radical as this shift proved, the standard remained objective. Whether Canterbury actually understood the risks did not matter; rather, what was required was to share the risks that a reasonable physician might anticipate that a reasonable patient might wish to know in a manner that such a reasonable patient could understand. The case proved transformative and has since been widely – albeit not universally – followed.[147] Although courts in Great Britain proved more resistant to this approach, rejecting the "prudent patient" standard as recently as *Sidaway v. Bethlem Royal Hospital* (1985),[148] more recent decisions, most notably *Montgomery v. Lanarkshire Health Board* (2015), have brought the United Kingdom in line with the United States and much of the commonwealth.[149] Unfortunately, the ruling did Jerry Canterbury himself little good, as the case was remanded for trial, and the jury determined that Dr. Spence had indeed met this standard.

3.2.4 Alternative Interventions

Early discussions of informed consent – both in the courts and in the ethics literature – largely focused upon the specific treatment being offered by the physician, rather than any potential remedies not being offered. In fact, "the claim that patients could have avoided harm if only they had been told about … the available alternatives usually carrie[d] little persuasive weight."[150] That approach placed the burden upon patients to ascertain the best possible treatments on their own – either through research or second opinions. Yet in a case as infamous in psychiatry as *Helling* has become in general medicine, *Osheroff v. Chestnut Lodge*,[151] a psychiatric patient named Rafael "Ray" Osheroff asserted that physicians did have an affirmative duty to inform patients of alternative treatment options. In that Maryland case, the plaintiff – himself a practicing nephrologist – was admitted to one of the nation's most prestigious private psychiatric facilities, the venerable Chestnut Lodge, in 1979, suffering from crippling depression and anxiety.[152] For seven months, he received only talk therapy, rather than antidepressant medication, with no significant improvement. His family then removed him from Chestnut Lodge and admitted him to another upscale facility, Connecticut's Silver Hill, where he showed improvement after only three weeks on a combination of phenothiazines and tricyclic antidepressants.[153] Osheroff returned to the practice of medicine within the year. He then sued, claiming that Chestnut Lodge had failed to inform him that an alternative remedy to psychotherapy was possible. Although the case was ultimately settled out of court, the nature of the claim and the hospital's decision to settle generated considerable debate. The result was a significant shift in medical practice and culture. At present, physicians are generally expected to inform patients of alternative reasonable treatment options. That does not mean that every psychoanalyst must offer pharmacological therapy, only that a therapist treating a patient for depression or anxiety should, at a minimum, inform a patient that additional effective interventions may be available from other providers. Documenting such a warning may prove essential in avoiding liability.

3.2.5 Limitations

Informed consent generally requires both providing information to patients and then honoring their choices, but the right is not without its limitations. First, a general consensus exists that physicians do not have a duty to offer patients options that are not anticipated to be therapeutic or are likely to prove futile. While defining medical futility is a complex matter well beyond the scope of this paper, once a determination has been made that such an intervention is truly futile, most commentators agree that providers need present it as an option to patients.[154,155]

Other situations arise in which providing a patient with all relevant information for an informed choice is not sufficient to avoid liability. The most well known of these are circumstances in which a physician and patient mutually agree to contract outside the standard of care. For instance, if a surgeon says, "Usually I operate with a scalpel, but for half price, I will operate with a cleaver," and a negative outcome results, the physician will generally be responsible for any resulting injuries – notwithstanding that the patient had a complete understanding of the risks and fully agreed to accept them. Obviously, few surgeons offer to operate with substandard equipment. However, in areas likely alternative and complementary medicine, attempts to contract outside the standard of care are surprisingly commonplace, and informed consent may not prove an adequate defense. If a treatment is "supported by a considerable number of respected and recognized medical experts," the physician may still be protected under what is known as the "two school of thought" or "respected minority" doctrine, even if the intervention does not fall within the standard of care.[156,157] However, if a physician is the first, or one of only a few, providers engaged in a particular course of treatment, informed consent will not absolve him of liability. Instead, participation in a clinical research trial – with appropriate safeguards – is the only way to avoid legal responsibility for poor results.

Autonomy does also not afford a patient the right to micromanage care when doing so contravenes established medical norms or is likely to disrupt clinical practice. So, while a capacitated patient has the right to decide on a course of treatment, and to know the risks and benefits of such treatment, the physician can refuse to defer to the patient with regard to certain decisions that might be thought of as "ministerial." For instance, if a surgeon generally performs a particular operation with a four-inch Gosset retractor, a request by a patient that a five-inch retractor be used instead need not be respected – even in an emergency – without some meaningful and relevant justification for the patient's demand. Similarly, the operating room schedule need not be rearranged to meet the preferences of a patient for an earlier or later procedure. While autonomy means the patient is often right, he is not always so. Of course, a capacitated patient is always welcome to turn down care in such situations, even placing his life at risk, but a physician will generally not be held liable for failing to accommodate such preferences.

3.3 The Right not to Inform

The duty of informed consent is not without its exceptions. Philosopher Onora Oneil has saliently observed that "informed consent cannot be relevant to all medical decisions, because it cannot be provided by patients who are incompetent to consent, cannot be used in choosing public health policies, cannot be secured for all disclosure of third party information, and cannot be obtained from those who are vulnerable or dependent."[158] In certain contexts, even generally capable patients may lose the ability to consent meaningfully, such as when they become "emotionally overwhelmed" by information or this information becomes too complex process.[159] Under such circumstances, the responsible clinician may choose to postpone the consent process, if possible, or strive to simplify the choice for the patient. Common sense suggests that informed consent not be required in certain circumstances, such as emergencies, when the acuity of the situation may prevent a meaningful consent process. Requiring informed consent prior to performing an emergent Heimlich maneuver on a choking patient, for instance, would likely defeat the entire purpose of the intervention. Similarly, most commentators accept waiver by the capacitated patient as grounds for overriding the general requirement, as the right *not* to obtain information also advances the underlying principle of autonomy.[160] In addition, "generally known risks" that do not require any medical training or expertise are often also exempt.[161] Several other situations prove more challenging to clinicians and ethicists. What follows is a discussion of three areas in which some, but not all, jurisdictions have accepted carve-outs to the general requirement that capacitated patients give informed consent. These situations include those involving therapeutic privilege, those in which such disclosures run contrary to the patient's purported cultural or religious values, and those in which the physician chooses not to share the comparatively better skills of competing providers.

3.3.1 Therapeutic Privilege

The doctrine of "therapeutic privilege" permits the "withholding information by a clinician, with the underlying notion that the disclosure of this information would inflict harm or suffering upon the patient."[162] The purpose of this exception to the rule of informed consent is "to avoid causing harm to the patient by disclosure, not to avoid a patient's refusal of treatment"[163]; it is intended to be invoked sparingly. Yet as Margaret Somerville has pointed out, even determining how to ascertain whether harm will occur is problematic. Under different approaches, she noted, "therapeutic privilege may apply when the reasonable physician in the same circumstances would have thought that the reasonable patient would be harmed by receiving the information normally required to be disclosed" or "when the reasonable physician would have thought that this particular patient would be harmed" or "when this particular physician would have thought that the reasonable" or "when this particular physician would have thought that this particular patient would be harmed."[164] Unfortunately, no clear "guidelines exist for implementing therapeutic privilege" and it is rarely upheld by the courts.[165,166] In the United States, the privilege is generally confined to extreme cases: Specifically, when divulging particular information to a specific patients – based upon his known history – is highly likely to cause him specific, significant harm. For instance, a physician might postpone sharing a grave diagnosis with a patient, even if doing so involved some prevarication, in order to ensure that appropriate social supports are present before disclosure, if when last informed of a serious diagnosis, the patient attempted to harm himself. Short-term delays in revealing physician intent are frequently used in the setting of involuntary psychiatric admission in order to protect patient safety.

What is clear is that the privilege does not extend to general beliefs that many patients will benefit from a more optimistic assessment of their illness than the facts indicate. Clearly, in the United States, one cannot withhold a diagnosis of cancer or heart disease on the grounds that patients who believe they have a better prognosis may enjoy higher qualities of life. Yet other jurisdictions have accepted a somewhat broader approach to therapeutic privilege. For instance, one Australian province allows the invocation of the privilege if disclosure will lead to a worse outcome based upon "the patient's personality," "temperament" or "attitude."[167] Yet the trend in common-law countries over the past several decades has been toward full disclosure, so the safest course is to err on the side of transparency. Some common law courts outside the United States have rejected the principle of therapeutic privilege entirely in favor of absolute truth telling. Cases that follow this reasoning include *Meyers Estate et al. v. Rogers* (Ontario, 1991), *Pittman Estate v. Bain* (Ontario, 1994), *Castell v. De Greef* (South Africa, 1994) and *Teik Huat Tai v. Saxon* (Western Australia, 1996)[168] Ama Kyerewaa Edwin, a critic of the practice, has argued that, "While patients may not have the medical expertise of their doctors, they are nonetheless in a better position to determine what is in their best interest based on the information made available to them."[169] Needless to say, deceiving patients – even temporarily and for a good reason – can damage the therapeutic relationship.

3.3.2 Respect for Values

Waiver has long been recognized as an exception the principle of informed consent. If a patient does not wish to know about his medical condition, such knowledge cannot ethically be forced upon him. A more complex situation arises when a family member attempts to waive such informed consent on behalf of a patient without that patient's knowledge. In general, with a capacitated patients, such waivers are invalid. Yet a relative may assert that the patient would not want to know, complicating this issue. Such requests often occur when the patient belongs to a cultural or religious community which still ascribes to paternalistic notions of medical care and/or where disclosing risks or a poor prognosis may even be interpreted by a patient as the physician not caring about the patient's welfare. What is to be done when a grandchild informs an oncologist, "In our culture, we don't tell our grandparents that they have cancer"?[170] The difficulty here is that one cannot ask the patient whether or not he would want to be informed if he had cancer, because asking the question telegraphs the answer. By analogy, imagine asking one's spouse, "If I were having an extramarital affair, would you want me to tell you?" At the same time, the grandchild might be mistaken about the grandparent's wishes, which in fact might not actually align with the broader cultural or religious practices of their community. Due diligence in such case is required. First, the physician must search for prior evidence regarding whether the patient might indeed wish to waive informed consent in such circumstances – either through written or oral indications, or through decisions the patient has made in relation to his own loved ones. For instance, in taking a routine family history, a physician might inquire if any of the patient's relatives have died of cancer, and then inquire whether such a relative had been informed of the diagnosis. If the current patient says no, that might open an opportunity to discuss the patient's own wishes without raising suspicions.

Another complicating factor in such cases is that distinct principles may apply in slightly varied situations. For instance, the best approach may differ in cases where the diagnosis is clear-cut and no medical decisions must occur from cases that require decisions about future treatment and goals of care. Of course, as noted earlier, even when no medical decisions must be made, a dying patient might wish to make non-medical choices such as rewriting a will or attempting to reconcile with estranged family members. Since honoring a vicarious waiver will deny the patient such opportunities, care must be taken to ensure that the waiver is actually consistent with the patient's overall wishes, an assessment that may be extraordinarily difficult to operationalize with adequate confidence in practice.

3.3.3 The Skills of Competitors

A third controversial exception to the general principle of informed consent involves the disclosure of the ability of a physician's competitors to perform the intervention more effectively. Note that this differs from the question that arose in *Osheroff v. Chestnut Lodge*: The question in that case was whether more effective treatment modalities were available, not whether other providers were more skilled in administering those remedies. As a general rule, neither courts nor ethical canons have required physicians to compare their own skills to those of others. For instance, one can imagine a neurosurgeon at a small community hospital who has a 50 percent success rate in clipping brain aneurysms. If a tertiary care center an hour away by care has a 75 percent survival rate, the first surgeons is not expected to share this information with a prospective patient – although he may certainly choose to do so. This exception applies even if the patient has ample opportunity to drive to the tertiary care center. The rationale behind this rule is not merely to shield less skilled providers from malpractice claims. Rather, the approach serves the function of furthering equity. If providers were compelled to compare their own skills to those of others, many patients would shift care to more skilled clinicians – creating bottlenecks for treatment as patients swamp talented physicians, while denying less talented physicians an opportunity to improve their abilities. As a result, those with higher success rates would become even more able, while those with lower success rates would see their abilities decline from lack of practice. The consequence of these trends would have significant implications for patients with limited resources and social capital – as they may prove less table to travel for care. Informing a patient at a small community hospital that the success rate for a procedure is higher at Massachusetts General Hospital or the Mayo Clinic may be of no more practical value than informing her that the success rate is higher on the moon. In addition, emphasizing to such patients that better care exists, when they cannot access it, may prove demoralizing.

A more complex challenge arises when a patient asks directly: "Can anyone else provide better care?" In such circumstances, overtly deceiving the patients is clearly unethical, whether or not doing so might prove a basis for a malpractice claim. Less clear is whether this ethical challenge is resolved if the physician informs the patient that she cannot comment on the skills of other providers. Doing so may be a form of deception by omission, especially if the physician knows that other clinicians have more successful outcomes. At the same time, requiring full disclosure only if asked raises equity concerns: namely, that only those patients who possess enough health care knowledge to know to inquire, and also feel empowered to do so, will benefit from such comparative information. Finally, withholding information regarding comparative talents may generally be permissible, but at the extremes, may prove ethically unacceptable. A physician who has *never* before performed a particular intervention, for instance, might be reasonably expected to share this information with a patient. Similarly, patients will generally want to know if their care is being provided by a house officer or a trainee. Determining precisely where routine differentials in ability or experience transcend into extremes remains an open and highly fraught question.

4 Advance Directives and Third-Party Decision-Making

When a patient is determined to lack capacity to make medical decisions, how decisions should be made on that patient's behalf has long challenged medical professionals. Prior to the 1960s and the rise of an ethical system in the West that placed primacy on patient autonomy, doctors often rendered such decisions on their own, with varying degrees of input from patients' families. Such an approach proved consistent with more general medical practice in which even patients who possessed full decisional capacity often had healthcare information withheld from them and frequently had decisions rendered on their behalf. However, as capacitated patients gained more control over their own healthcare, calls arose to grant similar authority to patients lacking capacity – either through advance directives or through third-party decision-making.

4.1 Historical Background

Various movements that favored voluntary euthanasia or assistance in dying for the critically ill emerged in the United States as early as the first decade of the twentieth century. Cincinnati heiress Anna S. Hall (1976–1924) became an early pioneer in such efforts, her movement particularly concerned with those suffering from severe injuries as a result of fires and railway accidents.[171] Although the legacy of involuntary euthanasia in Nazi Germany cast a dark shadow over such efforts, they continued to appear in the academic literature, particularly the work of British legal scholar Glanville Williams, who advocated for the "establishment of a means to immunize relatives or physicians who would administer a means of ending life upon a patient who is suffering great pain from an incurable disease for which there is no cure or relief and which is fatal."[172,173] Both the Euthanasia Society (now Dying with Dignity) in England and the Euthanasia Society of America proposed similar protocols for operationalizing his views with various safeguards including reviews by doctors. Neither proposal gained widespread traction. Under both common law and statutory criminal law in the United States and in the United Kingdom, those who abetted suicide or voluntary euthanasia faced serious legal repercussions. One particular consequence of these laws was that even when a patient could not be subjected to treatment without his consent, if he were "in a condition in which his consent [could not] be expressed, the physician [had to] assume that the patient wishe[d] to be treated to preserve his life."[174] To address this challenge, American human rights activist Luis Kutner (1908–1993) first proposed in a controversial article in the *Indiana Law Journal* in 1968 that, when a patient signs an informed consent document for "surgery or other radical treatment," the patient, "while still retaining his mental faculties and the ability to convey his thoughts, could append to such a document a clause providing that, if his condition becomes incurable and his bodily state vegetative with no possibility that he could recover his complete faculties, his consent to further treatment would be terminated."[175] As Kutner later described his proposal, "When an individual patient has no desire to be kept in a state of complete and indefinite vegetated animation with no possibility of recovering his mental and physical faculties, that individual, while still in control of all his faculties and his ability to express himself, could still retain the right of privacy over his body in determining whether he should be permitted to die by way of a document."[176] Kutner suggested various names for such a document including "living will," "declaration determining the termination of life," "testament permitting death," "declaration for bodily autonomy," "declaration for ending treatment," and "body trust," but the first of these terms – living will – acquired widespread acceptance.[177] Kutner argued that an "individual could carry this document on his person at all times, while his wife, physician, lawyer, or confidant would have the original copy" and that "personal possession of the document would create a strong presumption that one regards it as binding."[178] Five years later, upon Kutner's recommendation, the American Society for Euthanasia produced the first printed living wills. These "short and straightforward" documents stated that "if an individual had no chance of a reasonable recovery, he or she should be allowed to die."[179] Yet at the time, these documents had no legal validity in any jurisdiction.

A far more challenging case arose in situations where patients had not previously executed a document expressing their wishes for life preservation if rendered permanently incapacitated. This issue became the subject of national headlines, and public debate, as a result of the tragedy of a 21-year-old New

Jersey woman, Karen Ann Quinlan. Quinlan had ingested diazepam and alcohol while on an extreme diet that left her in a persistent vegetative state (PVS). Her physicians believed that she would require "ventilator support and tube feedings for the rest of her life."[180] Her parents, Joseph and Julia, sought to have the ventilator turned off, but Saint Clare's Hospital refused. In the parents' subsequent legal challenge, the Supreme Court of the State of New Jersey ultimately ruled in 1976 that Joseph Quinlan, as Karen's court-appointed guardian, might terminate her ventilator support – at least under certain circumstances. As Justice Richard Hughes wrote for the court:

> "We repeat for the sake of emphasis and clarity that upon the concurrence of the guardian and family of Karen, should the responsible attending physicians conclude that there is no reasonable possibility of Karen's ever emerging from her present comatose condition to a cognitive, sapient state and that the life-support apparatus now being administered to Karen should be discontinued, they shall consult with the hospital "Ethics Committee" or like body of the institution in which Karen is then hospitalized" and "if that consultative body agrees that there is no reasonable possibility of Karen's ever emerging from her present comatose condition to a cognitive, sapient state, the present life-support system may be withdrawn and said action shall be without any civil or criminal liability therefor, on the part of any participant, whether guardian, physician, hospital or others."[181]

Shortly thereafter, California became the first state to incorporate the validity of advance directives into law with the recognition of living wills through the Natural Death Act of 1976.[182] In the ensuing two decades, more than forty American states adopted some form of recognition of advance directives, although with considerable variability in their scope.[183] This evolution culminated in federal action in 1990 with the passage of the Patient Self Determination Act. This statute required that hospitalized patients be "asked about advanced directives," that hospital staff "document any wishes the patient might have with regard to the care they want or do not want," and, most importantly, "mandate[d] that patient advance directives be implemented if necessary, assuming those wishes are legally valid and permissible by State law."[184] At present, all American jurisdictions allow for patients to designate agents through advance directives.

4.2 Agents: Classifications and Duties

In the absence of a living will or similar document, American jurisdictions now generally allow third-party decision-making for patients. However, this has not always been the case; even many early living will statutes made "no provision for the person to designate another person to make decisions on his or her behalf or set forth the criteria for such decisions."[185] The first statutes establishing such third-party agents, such as California's Durable Power of Attorney for Health Care (1984), were generally adopted later than those authorizing living wills.[186] More recently, some states have adopted both forms of advance directives concurrently. In Great Britain, decisions for incapacitated patients fell under the jurisdiction of the courts until the adoption of the Mental Health Act of 1959, after which for many years the responsibility became that of medical professionals.[187] A series of cases, culminating in *Malette v. Shulman* (1992), established the principle that "prior refusals of medical treatment would be binding provided" that "at the time the patient made the declaration he must have been competent to consent or refuse the treatment," that "the patient must have anticipated and intended his decision to apply to the circumstances that ultimately prevail" and that "patient's decision must have been reached without undue influence."[188] In 2005, the right to appoint a proxy in England was finally codified into statute in the Mental Capacity Act.

The terminology that governs third-party agents is complex and the same role is often referred to by different names in different jurisdictions. Yet two broad categories of third-party decision-makers exist: appointed agents and default agents. (Guardians or conservators, appointed by the courts to render *all* decisions for patients, generally have authority over healthcare choices by nature of the individual's legally determined incompetence, and are not considered here.) Appointed agents are individuals previously designated by patients, when in possession of capacity to do so, to make healthcare decisions on their behalf once they become incapacitated. The rules governing whether such appointments may be oral or must be written, and whether they must be witnessed, vary between states. Appointed agents may be variously known as "health care proxies" (HCPs), "medical powers of attorney" (MPOAs), or "health care powers of attorney" (HCPOAs); some jurisdictions add the descriptor "durable" to MPOAs and HCPOAs to distinguish them from temporary agents. Other terms found in state statutes include "healthcare agent," "healthcare representative," "attorney-in-fact" and "patient advocate."[189] Appointed agents can operate in conjunction with living wills. In some jurisdictions, a living will may trump the wishes expressed by a proxy in areas where there is disagreement; on others, the preference that best effectuates the wishes of the incapacitated patient governs – requiring an assessment that may depend upon whether the patient countermanded part or all of a living will to the proxy or other parties.

An individual must have capacity to appoint a third-party agent, but most state statutes do not provide any specific standards for doing so. In two states, Utah and Vermont, statutes define specific standards that differ notably from those required for other forms of healthcare decision-making. In Vermont, the statute requires the appointing party to possess "a basic understanding of what it means to have another individual make healthcare decisions for oneself" and of "who would be an appropriate individual to make those decisions."[190] Utah's statute is far less precise, requiring only that the appointing party "understands the consequences of appointing a particular person as agent" with a rebuttable presumption of capacity to make the appointment.[191] In other jurisdictions, considerable latitude may exist whether to use the same standards that are required for decisional capacity or a lower standard, largely depending upon case law and clinical practice.[192] Individuals are generally permitted to appoint almost any competent adult as a health care proxy, although most jurisdictions exclude current medical providers – unless those providers are also relatives and recuse themselves from ongoing involvement in providing medical care to the individual they represent. It should be emphasized that the capacity *to appoint* a healthcare agent is not the same as the capacity *of* the healthcare agent. Assessment of the agent's capacity, if necessary, is generally not handled by consult-liaison psychiatrists, but requires evaluation by an individual with formal forensic training and usually occurs in conjunction with a legal proceeding to appoint an alternative decision-maker. Similarly, surrogates may possess cultural or religious values that physicians find "unreasonable" or even "offensive."[193] How to handle such surrogates usually falls to the courts, although some evidence suggests that mediation may also prove effective.[194]

Individuals appointing third-party decision-makers should be encouraged to inform these third parties of their wishes in advance, especially in jurisdictions that use a standard based upon the patient's own preferences. However, considerable data shows non-concordance between patients' preferences and third-party agents' understanding of patients' preferences, although agents do show higher concordance rates than patients' physicians.[195–196] As Fredric Wolinsky et al. have noted in a recent study, "concordance levels" between patient's wishes and those reported by their spouse proxies "are disappointing" and "suggest substantial disagreements between husbands' and wives' responses for themselves compared to their spouses' proxy responses on their behalf."[197] One cause for this disparity may be a failure to discuss preferences with agents during times of good health. Allison Seckler et al., for instance, found that only 16 percent of patients ever discussed their wishes regarding cardiac resuscitation with their third-party decision-makers.[198] However, even with these shortcomings, agents do demonstrate "imperfect accuracy" in matching patient preferences that still remains substantially better than chance.[199]

Of course, that does not indicate that appointed agents are not valuable in healthcare, but rather that physicians must do a better job of educating patients that the appointment of an agent is the beginning of an ongoing process of sharing one's wishes with one's agent, not a concluding decision that requires no further engagement. Another concern of note is that structural bias continues to reduce the rates at which Black and Hispanic individuals appoint third-party agents.[200] However, many of these systemic factors, such as "knowledge of health care proxies" and "beliefs about the necessity of a formally appointed health care agent in the presence of involved family" are likely modifiable with appropriate engagement by physicians.[201]

Many individuals do not appoint any third-party agent in advance. In the absence of such an appointed proxy or power-of-attorney, decision-making authority historically fell to either the courts or to medical providers. In 1991, the American Medical Association (AMA) issued a report, "Decisions to Forgo Life-Sustaining Treatment for Incompetent Patients," that proposed a role for third-parties in decision-making, even in the absence of prior appointment by the incapacitated patients. One by one, states adopted laws that permitted such default agents – generally known as surrogates – to make some or all decisions for incapacitated patients. As of 2023, only four American states continue to lack such statutes: Massachusetts, Missouri, Minnesota, and Rhode Island. In other jurisdictions, the powers of such surrogates vary considerably. In some jurisdictions, they have the same powers as appointed agents, while in others, their powers are curtailed – especially with regard to withdrawing life support or artificial nutrition. In New York State, for instance, proxies may terminate life support under any circumstances which are in accordance with the patient's known wishes, while surrogates are limited to withdrawing such care under defined conditions. Another key advantage of appointed agents over default surrogates is that proxy forms are generally recognized across state lines, assuming that the original health care proxy meets the requirements of the second state. Many countries recognize foreign health care proxy forms as well. In contrast, the recognition of surrogates largely depends upon the state of current jurisdiction. Finally, the concept of a HPC or HCPOA is familiar to most physicians and increases the likelihood that medical providers will consult third-party agents. As default surrogacy is a more recent phenomenon, some providers may prove less familiar with the concept and, in emergency circumstances, may even inappropriately override such default agents. In light of these considerations, outpatient physicians should make every possible effort to educate patients regarding the benefits of executing advance directives.

New York State's Family Healthcare Decisions Act offers a fairly typical example of the order of succession enumerated in many state statutes; surrogacy falls to individuals in the following order 1) "A guardian authorized to decide about health care"; 2) "the spouse, if not legally separated from the patient, or the domestic partner"; "a son or daughter eighteen years of age or older"; "a parent"; "a brother or sister eighteen years of age or older," and "a close friend." Patients should be aware that although "common law" marriage is legal in some jurisdictions, and is frequently references in popular culture, many states do not recognize common law marriage, which is a phenomenon distinct from domestic partnership, and individuals may learn too late that their so-called "common law" spouse has no decisional authority under their state's surrogacy laws, as is the case in New York.[202]

Among the issues that have challenged lawmakers in various jurisdictions with regard to surrogacy statutes are whether and how to recognize same-sex partners, especially before gay marriage was legalized cross the United States; the role of partners or other agents in terminating life-support for pregnant individuals; and how to resolve conflict when disagreements arise among surrogate decision-makers of equal rank.[203,204] For instance, a incapacitated patient may have two sisters, both sincere in their beliefs, one of whom contends that a patient would have wanted life support continued and the other who insists that the patient would have wanted it withdrawn. States that follow the Uniform Health-Care Decisions Act follow the principle that if disagreement exists among a class of surrogates that cannot be resolved, "the supervising health-care provider shall comply with the decision of a majority of the members of that class who have communicated their views to the provider." Yet seven states require "a consensus . . . among equal-priority surrogates," while one gives doctors the authority to select the equal-rank party who "appears to be the best qualified."[205] Finally, it should be noted that "next of kin" and surrogate are not synonymous terms in all jurisdictions, although they are used interchangeably on occasion, and each of these parties may gain decision-making authority under different circumstances and with different parameters. As a general rule, one's "next of kin" is one's closest legal relation and gains authority over one's property after death in cases of intestacy. "Next of kin" rules, either statutory or via common law, may incorporate distant relations. In contrast, surrogacy statutes generally do not incorporate such distant blood relatives, but often do have provisions for close friends. This discrepancy stems from the distinct purposes of each law: "Next of kin" laws are designed to ensure the passage of property from higher generations to lower generations and to guarantee certainty of ownership; in contrast, surrogacy laws are intended to ensure that the party who knows the patient best renders the patient's healthcare decisions.

In some cases, incapacitated patients in need of medical care will have neither an advance directive nor a default surrogate. These individuals have historically been known by a range of terms including "unbefriended adults" and "adult orphans," although both of these descriptors are increasingly considered stigmatized. Agentless patients, whose ranks often include the elderly and those with limited social capital, pose distinct challenges for the healthcare system.[206] Unfortunately, as legal scholar Thaddeus Mason Pope observes, "no consensus" exists regarding how such cases should be handled and, [a]cross the United States, few jurisdictions have developed laws or policies that adequately protect this most vulnerable population."[207] Various professional organizations, most notably the American Geriatrics Society (AGS), have issued model guidelines for such cases with the aim of helping to "decrease state-to-state variability in legal standards," but so far these efforts have proven largely ineffective.[208] In emergencies, a broad consensus has developed that life-saving care should be provided to such patients, but much more variability exists in non-emergent circumstances. Some states, including Florida, Illinois, Kentucky, and Virginia, use the guardianship process for such patients in non-emergent situations, while others, such as North Carolina, Alabama, Arizona, Connecticut and North Carolina, "give broad statutory authorization for healthcare decision-making to the patient's attending physician."[209] The AGS urges clinicians to "ensure procedural fairness regarding medical decision-making by adopting a systematic, team-based approach that synthesizes all available evidence regarding unbefriended older adults' treatment preferences," but often the preferences of unrepresented patients cannot be determined.[210] In many states, guardians or medical providers are then called upon to act in the "best interests" of such patients. Yet what precisely determines the best interests of these individuals, particularly in matters of nuanced and complex medical decision-making, often defies easy resolution.

4.3 Decisional Standards

Once a determination has been made that a patient lacks capacity to render a particular healthcare decision and that a representative must do so, the next question that arises is what criterion the third-party agent should use in making a choice. The widely accepted standard today in the United States is "substituted judgment" or "vicarious judgement" in which agents "try to make the decision that the patient would have made if he or she were able to make decisions."[211] Dan Brock and Allen Buchanan have elaborated on this principle, requiring that an agent "choose as the patient would choose if the patient were competent and aware of both the medical options and of the facts about his or her condition, including the fact that he or she is incompetent."[212] This "substituted judgment" doctrine first developed in the nineteenth century in relation to distribution of an incapacitated person's property.[213] As early as 1969, in the case of *Strunk v. Strunk*, the principle was used to allow an incompetent man to donate a kidney to his brother.[214] In theory, the goal is to effectuate the patient's autonomy. However, substituted judgement is not without significant limitations. First, some patients have never previously expressed any preference on the subject; in many cases, medical situations arise which the average layperson could never even have anticipated or contemplated. In such cases, many clinicians choose to base decisions upon the rather amorphous evidence on how the patient lived his life. David Wendler and John Phillips have branded this method the "endorsed life" approach.[215] However, this principle is also problematic: How one chooses to live while in good health does not necessarily offer guidance regarding what choices one might make when ill. Second, patients may have expressed prior wishes that stand in conflict with each other. Thomas Gutheil and Paul Appelbaum offer the example of a patient "who repeatedly engages in self-destructive and self-defeating patterns of behavior" and then "seeks psychiatric help to learn how to stop this pattern from recurring"; if that patient becomes psychotic, and is declared incompetent, should the agent use the previous (competent) self-defeating, self-destructive pattern of living" to guide decisions or "the fact that the ward, when competent, sought help to overcome this very pattern, and thus might be expected again to seek help, this time in the form of treatment."[216] Third, "substituted judgment" fails to account for the well-documented "phenomenon of 'bargaining down', in which patients finding themselves in a compromised state may prove willing to accept life at a level of impairment that they previously had not thought they would wish to tolerate."[217,218] This frequent occurrence threatens to undermine the entire concept of what Ronald Dworkin calls "precedent autonomy," as the patient's previous autonomous wishes may not be relevant to the person in his current state.[219] Despite these shortcomings, the majority of American physicians continue to prefer a "substituted judgment" standard over alternatives.[220]

A different approach, the "best interest" standard, prioritizes the incapacitated patient's welfare over his preferences and is consistent with the bioethical principles of beneficence and nonmaleficence. Such an approach was widely used by both physicians and courts in the era before the development of advance directives and default surrogacy laws. However, the best interest standard "fails to protect the patient's right to self-determination," so has fallen out of favor.[221] A "best interest" approach remains preferred by some ethicists, and has been incorporated into the laws of many jurisdictions, in regard to a subset of cases for which "substituted judgment" is believed to be either poorly suited or logistically impossible. As discussed above, choices regarding adults without representation are one such category. A second set of cases in which a "best interest" standard is often applied are those in which the patient never, in the past, possessed the capacity to render the decision in question.[222] Patients with significant cognitive impairment from childhood, for instance, may require such an approach: Many states view it as a bridge too far to inquire what a patient who has an IQ of 20 might choose if she had an IQ of 120. Finally, decision-making for pediatric patients relies to some degree upon a best-interest standard, especially when high-stakes decisions threatening severe morbidity or mortality are involved.[223] However, in other cases, a hybrid model is used, with physicians and courts often looking to the patient's preferences and seeking "assent," even when formal legal consent is not possible.[224] Such an approach has been endorsed by the American Academy of Pediatrics.[225] As the age of the minor increases and the stakes of the decision decrease, the minor is often, in practice, afforded more say in medical choices – even if the legal definition of the age of majority remains a *per se* cutoff of eighteen or nineteen years.

4.4 Evidentiary Standards

In many cases, when trying to effectuate a patient's prior wishes, the question arises as to how strong or clear the evidence must be to conclude those preferences to be dispositive. This concern has historically arisen in situations in which third-party representatives wish to withdraw life-saving or life-prolonging care. Over the years, the American legal system has used different evidentiary standards to address different types of factual questions. At one extreme, criminal conviction generally requires the prosecution to advance proof of guilt "beyond reasonable doubt." The Federal Judicial Center defines this standard as being "proof that leaves you firmly convinced."[226] At the other extreme, many forms of civil litigation, such as contract disputes, merely require that one party establish its case by a "preponderance of the evidence," meaning more likely than not, or in mathematical terms, 50 percent + 1. In between these two poles, the standard of "clear and convincing evidence" has been used as the level of proof required in many high-stakes matters outside of criminal law such as civil commitment and revocation of child custody. The standard has been defined as requiring evidence "so clear as to leave no substantial doubt" and "sufficiently strong to command the unhesitating assent of every reasonable mind."[227] Quantifying such a standard numerically has proven challenging, but one approach takes "the preponderance of the evidence standard as equivalent to 0.5, proof beyond a reasonable doubt as roughly 0.95, and proof by clear and convincing evidence even more roughly . . . 0.75."[228] A 2016 survey of American federal judges conducted by Richard Seltzer and colleagues placed the average assessment of clear and convincing evidence at 73.4 percent certainty.[229] Over the past four decades, courts and legislatures have been called upon to determine which of these standards – or another entirely – should be the minimum required for decisions based upon substituted judgment.

The Constitutional question regarding the maximum amount of evidence a state might require before an authorized representative could turn down life-prolonging care for an incapacitated patient was addressed by the United States Supreme Court in 1990 in the high profile case of *Cruzan v. Director, Missouri Department of Health*.[230] Of note, the case was also the first time the Supreme Court had directly addressed issues surrounding the right to die. The human tragedy underlying the case was a 1983 automobile accident in Missouri that left a 25-year-old woman, Nancy Cruzan, in a persistent vegetative state (PVS) and on a feeding tube. Her parents, Lester and Joyce, petitioned the state court to terminate artificial nutrition, as they "firmly believe[d] she would not [have] want[ed] to have tube feeding continued under such circumstances, in part on the basis of her own statement that she would not want to continue to live if she could not be 'at least halfway normal.'"[231] While a lower court ruled in the family's favor, the State Supreme Court of Missouri reversed on the grounds that the Missouri living will statute preferred the preservation of life, and that under the law, "no person can assume that choice for an incompetent in the absence of ... clear and convincing, inherently reliable evidence."[232] The Cruzans then appealed their case to the United States Supreme Court. Writing for a conservative 5–4 majority, Chief Justice William Rehnquist found that "a state may apply a clear and convincing evidence standard in proceedings where a guardian seeks to discontinue nutrition and hydration of a person diagnosed to be in a persistent vegetative state."[233] In other words, states may demand "clear and convincing evidence" of the patient's preferences in cases requiring substituted judgement. With regard to the withdrawal of life-saving or life-preserving care, some, but not all, states have chosen to do so. Many jurisdictions allow more routine decisions to be made based upon a lower evidentiary standard.

The *Cruzan* decision has not been without its critics. Ethicist Susan Wolf has argued for a lower standard, contending that "a demand for so much formality and specificity, ultimately for the fulfilment of such impossible conditions," undermines the right to self-determination entirely, because it is inconsistent with "real life" and that people do not "speak about their own death in contract-talk, with all the i's dotted, terms stated, and ambiguities minimized."[234] In contrast, Constitutional law scholar Steven Calabresi has compared the withdrawal of life-sustaining care to other legal matters in which lives are at stake in arguing for an even higher standards. He writes that, "Ordinarily, in end-of-life cases such as those involving heinous murderers convicted of heinous crimes, we demand proof beyond a reasonable doubt before life may be ended," so a similar standard should apply in substituted judgment cases that may result in death.[235] The aftermath of the U. S. Supreme Court's decision in *Cruzan* raises another cautionary concern: that of so-called "good faith" perjury. In November 1990, Cruzan's parents presented additional evidence to the court, including testimony from co-workers that Cruzan had previously stated that she would never have wanted to live "like a vegetable," thereby meeting Missouri's clear and convincing evidence threshold.[236] These conversations may indeed have taken place, but that they had not emerged previously provoked suspicions. Speaking more generally, elevating the evidentiary standard may lead witnesses to attribute statements to the incapacitated patient – even when those precise statements did not occur – if, in doing so, they present testimony that does effectuate what they believe to be the patient's prior wishes. Needless to say, such a roundabout approach to a potentially accurate outcome undermines the rule of law and is far from ideal.

4.5 Mental Health

4.5.1 Psychiatric Advance Directives

Initial efforts to establish the moral and legal authority of advance directives generally excluded decision-making for patients with significant psychiatric illness. These de facto carve-outs are not surprising, as psychiatric and medical conditions have historically been treated very differently by the law. However, in the late 1970s and early 1980s, a pair of landmark cases in the United States began to reshape the legal landscape. In *Rennie v. Klein* (1978), the United States District Court for the District of New Jersey found that since "mental illness is not the equivalent of incompetency, which renders one incapable of giving informed consent to medical treatment," psychiatric patients who had not been declared incompetent had a limited right to reject medication and medical interventions.[237] In a similar case, *Rogers v. Commissioner of Mental Health* (1983), the Massachusetts Supreme Court found that "a committed mental patient is competent and has the right to make treatment decisions until the patient is adjudicated incompetent by a judge" – taking the matter out of physicians' hands entirely.[238] It was not until the following decade that psychiatric patients first attempted to guide prospective treatment after determinations of incapacity or incompetence. The first known instance occurred in 1992, when a Minnesota woman attempted to execute such a document.[239] Around this time, Minnesota also adopted the first known "psychiatric will" (or psychiatric advance directive/PAD) statute, which allowed for patients to *authorize* future treatment without need for a court order.[240] At the same time, the statute did not on the surface permit patients to prospectively *refuse* treatment once incapacitated. Three decades later, twenty-five states had adopted some form of legislation that either authorized PADs or incorporated psychiatric decision-making into existing advance directive statutes.[241] Similar statutes have also been adopted in Australia, New Zealand, Canada, and the Netherlands.

In most jurisdictions that permit PADs, they are either advisory or do not allow incompetent psychiatric patients to refuse treatment. One notable exception are those American states that fall into the Second Federal Judicial Circuit – Connecticut, New York, and Vermont. In a high profile 2003 case, *Hargrave v. Vermont,* the Second Circuit Court of Appeals upheld the right of a woman with schizophrenia, Nancy Hargrave, to authorize an appointed agent to refuse all psychiatric medication on her behalf, even if that meant permanent hospitalization on the grounds that she posed an ongoing threat to herself or others. The court found this document enforceable as a result of the Americans with Disabilities Act, "which requires that 'no qualified individual with a disability shall, by reason of such disability, be excluded from participation in or be denied the benefits of the services, programs, or activities of a public entity, or be subjected to discrimination by any such entity."[242] According to the court's reasoning, "having established a statutory basis for medical advance directives," a state "cannot exclude involuntarily committed psychiatric patients from its coverage."[243] No other courts in the United States have yet reached the same conclusion.

Binding psychiatric advance directives are a form of "Ulysses contract" intended to overcome *akrasia* or "weakness of the will."[244] The terminology stems from the Homeric epic, the *Odyssey,* in which the Greek King of Ithaca, Odysseus (Ulysses in Latin), wishes to hear the alluring music of the Sirens, lyrical, human-like creature, possibly mermaids, without approaching so close that he perishes upon the rocky shores that they inhabit. To avoid such a fate, Odysseus commands his sailors to bind him to a mast, while placing bees wax in their own ears, and to refuse to release him despite his gestures to the contrary while they pass the Sirens' lair. In other words, Odysseus overcomes his future *akrasia* with an irrevocable prospective order. Psychiatric advance directives offer psychiatric patients a similar opportunity to "commit themselves to a future treatment plan and [to] forfeit their own right to object to the decision(s) outlined in that plan, should they lose decision-making capacity" in the future.[245] Of course, binding oneself to a future outcome is a controversial proposition, especially in light of the "bargaining down" phenomenon discussed previously.

Psychiatric advance directives remain contentious. While concerns have been raised that such directives may "contain preferences that are unclear or incompatible with practice standards," a systematic review by Anne-Sophie

Gaillard and colleagues found PADs to "document clear, comprehensive, and clinically relevant preferences regarding future mental health crises."[246] Engaging with patients in the context of their own cultures has been shown to increase the efficacy of PADs.[247] Critics argue that some patients, even when they meet the minimum standards of capacity, lack a true appreciation of the consequences of binding themselves to remain psychotic in the future. Psychiatric advance directives are also not without financial implications for society at large, as patients who go untreated may pose a burden to the healthcare system and the taxpayers in the long run. When public will limits the overall resources allocated toward mental healthcare, the decision of one patient to refuse care may indirectly exhaust funds that could help other psychiatric patients in need.

4.5.2 Diminished Capacity

While the law recognizes that different types of healthcare decisions require differing levels of cognitive ability and understanding – a specific, sliding scale approach to capacity – it generally considers capacity for each specific decision on an all-or-none basis. Either a patient crosses the threshold of capacity for that decision, or he does not.[248] The reality in clinical care is that patients often suffer from some level of diminishment in their capacity due to outside forces, from duress to distress, that influence decision-making. "Diminished capacity" is a particularly relevant factor in patients who wish to turn down medical care as a result of depression; clinicians and courts must often decide whether the depression rises to such a level as to impair a medical decision rather than merely to influence it. In cases of patients with "poor medical prognoses and/or low quality of life" (PMP/LQL), the patient may suffer from depression "yet voice a desire for death under circumstances in which he would prefer death even if not depressed."[249] Whether agents should have authority to make decisions accordingly, or whether physicians and the courts should override the patient's prior wishes as a result of the depression, is a matter of ongoing debate. For instance, "a healthy person might well instruct a proxy, 'If I ever end up critically ill and depressed, please let me die rather than impose unwanted psychiatric care upon me, even if I have some prognosis for recovery'."[250] If that patient then loses capacity, but still expresses an interest in continuing life-preserving interventions, should the prior wishes as voiced by the agent trump the current wishes voiced by the incapacitated patient? Honoring "precedent autonomy" argues for upholding the prior wish; at the same time, terminating life-saving care for a patient who vocally expresses a wish for such treatment to continue may trouble many providers and ethicists. Yet if one does not uphold "precedent autonomy" in this instance, one is then hard pressed to find a distinction between this case and many others, risking the unravelling of the premise upon which the entire concept of advance directives is based.

4.5.3 Medical Aid in Dying

The issue of diminished capacity among certain psychiatric patients raises distinctive issues related to so-called "assisted suicide" or "medical aid-in-dying" (MAID), as this practice becomes legalized in an increasing number of jurisdictions. Most patients seeking MAID are terminally ill or suffering intractably, so the question inevitably arises whether these factors inherently cloud the judgment of those seeking to end their own lives. These issues are exacerbated when the patient's suffering is the result of psychiatric illness. In 1998, an unidentified, 53-year-old Swiss citizen with a history of bipolar disorder sought a prescription for a lethal dose of sodium pentobarbital, claiming "a right to self-determination under Article 8 of the European Convention on Human Rights," and a Lausanne court agreed.[251] Since that time, MAID for patients with mental illness had been legalized in several other nations including the Netherlands, Belgium, Luxembourg, and Canada.[252] These jurisdictions generally require multiple safeguards in such cases not demanded in general medical decision-making. The Netherlands and Belgium also allow certain patients, such as those suffering from progressive dementia, to authorize MAID at a future point, when they are no longer capable of making such decisions. Thus, a patient may be able to live with dignity until a future moment in time at which their present-day self would no longer wish to live, preventing people from being forced into MAID prematurely. Yet whether an agent or third-party ought to be permitted to authorize MAID on a patient's behalf, barring explicit prior authorization, remains an unsettled – and unsettling – ethical question.

4.6 Ongoing Challenges

4.6.1 Standards for "Never-Capacitated" Patients

Two state court cases from the late 1970s and early 1980s, one from New York and the other from Massachusetts, offered two distinct pathways for handling third-party decision-making for patients who, due to cognitive impairment, had never possessed capacity to render the medical decision in question. In the first of these cases, *Superintendent of Belchertown State School v. Saikewicz* (1977), involved a 67-year-old, institutionalized, non-verbal man with an I.Q. of 10 who had been diagnosed with acute myeloblastic monocytic leukemia (AMoL).[253]

He had no family willing to provide any relevant history or assist in medical decision-making. The Massachusetts courts had to decide whether to authorize chemotherapy that had the potential to extend Joseph Saikewicz's life at the expense of side effects. The Supreme Judicial Court of Massachusetts chose to impose a substituted judgment standard. The courts based their choice upon "that which would be made by the incompetent person, if that person were competent, but taking into account the present and future incompetency of the individual as one of the factors which would necessarily enter into the decision-making process of the competent person."[254] Critics including Daniel Lee objected on the grounds that "to speak of a right of refusal when no capacity for choice has ever existed, to make decisions in the name of autonomy when no authorization to do so has been extended by the patient, and to ascribe preferences to a patient when there is no track record on which to base these preferences is to over-extend the notion of patient autonomy and the accompanying notion of the right to refuse medical."[255] Most states subsequently rejected a substituted judgment standard for the so-called "never-capacitated," preferring a best-interest approach. The first of these cases, that of *In re Storar*, laid the groundwork for most of the legislation and case law that followed. A fifty-two-year-old New York man who had been cognitively impaired from birth and institutionalized for forty-seven years developed bladder cancer, for which his elderly mother sought to turn down care on his behalf.[256] In distinguishing the case from those of patients who had previously possessed capacity, where the New York courts required "clear and convincing evidence," Chief Justice Sol Wachtler wrote for a Court of Appeals majority that in the case of a never-capacitated patient like John Storar, it would be as "unrealistic to attempt to determine whether he would want to continue potentially life prolonging treatment if he were competent" as to ask whether, "if it snowed all summer would it then be winter?"[257] Instead, Wachtler used a best interest standard, which is now widely followed in most American jurisdictions.

Early challenges to the "best interest" approach in such cases argued that more deference ought to be afforded the wishes of families.[258] More recently, criticism of using the best interest standard has raised concerns regarding the ways in which doing so undermines the interests of cultural or religious minorities. Instead, these critics favor allowing third-party representatives to decide such cases based upon the values of the community into which the never-capacitated patient was born – rather than secular society at large. For instance, Appel has argued that "while not everyone shares the religious or cultural values of their parents or families, individuals usually do so at least until they willfully choose another set of values or traditions," influenced by both biological and environmental factors.[259] As a result, "since never-capacitated patients, if they had possessed capacity, would likely have experienced both the biological and environmental influences of their families, it follows that their preferences would have been much more likely to reflect those values than to approximate those of a hypothetical average or reasonable person."[260] Moreover, in such insular communities, "these values are likely to be well thought out and deeply held."[261] Allowing members of such groups like Chasidic Jews and Amish Mennonites to follow the community best-interest standard, rather than the broader societal one, seems to more closely approximate the goals of maximizing autonomy. Finally, one might not unreasonably conclude that such patients, if they did possess capacity, would wish to minimize distress to their families, giving them even further incentive to adhere to their community's standards. While such deference to community values has not yet eclipsed the traditional best interest approach in courts or legislatures, it continues to attract interest and support among clinicians and ethicists.

4.6.2 Absent Agents

One challenge that frequency arises in the hospital setting involves the location and accurate identification of the appropriate representative. In many circumstances, the patient will arrive in an acute medical crisis without the ability to communicate the identity of, or contact information for, an appropriate decision-maker. Even when such information is available, that does not mean that the representative will prove easy to find. Inevitably, clinicians must have guidance regarding how much effort is required to track down an agent before deciding that such an endeavor has failed, enabling them to stop searching and permanently circumvent that party for the next identifiable agent. Of note, while patients may designate joint agents and/or back-up agents, without specific prior authorization from the patient, representatives generally cannot designate alternative third parties as agents ahead of those of next rank (whether by appointment or default). For instance, if one's spouse is one's unappointed surrogate, but declines the role, the next decision maker will be the next party on the surrogacy list, such as an adult child; the spouse cannot assign a different party to render decisions. Most jurisdictions offer some statutory guidance on the level of exertion required to locate agents, such as a "reasonable" or "diligent" effort.[262] Such attempts should be tailored to the amount of time available before decisions must be rendered; as a general rule, if a higher-ranked agent presents later in the care process, that individual automatically assumes their rightful role as agent.

A related challenge involves "imposter" agents: third parties who identify themselves as representatives when they in fact do not have legal authority.

In many cases, these so-called "impostors" are acting in good faith – such as a "common law" spouse who presents as the patient's husband or wife without noting that no legal marriage has occurred. In other cases, purported representatives may act out of convenience, such as the close friend who claims to be a relative for a patient without known relations, believing that doing so will facilitate care. In extreme cases, individuals acting in bad faith will claim relationships or even forge healthcare proxy forms for nefarious purposes. For instance, a subletting roommate of a patient who suffers a sudden cognitive decline may fear eviction if a patient expires or is placed in a nursing home, so will claim to be an agent to keep the patient alive and at home – to prevent his own eviction. In most cases, the identity of the agent will prove apparent, such as when the patient confirms a familial relationship or friendship.[263] Physicians need not engage in invasive queries that risk undermining the therapeutic relationship, such as requesting a marriage license. However, if suspicions do arise, then the care team has an ethical obligation to engage in meaningful due diligence to confirm the purported agent has legal authority. Doing so ensures that both the autonomy and welfare of the incapacitated patient are best served.

4.6.3 Pregnant Patients

Third-party decision-making for incapacitated pregnant patients is among the most legally and ethically fraught issues in the field of bioethics. In many cases, of course, no conflict arises. For instance, when the healthcare interests of the patient and fetus coincide, and the agent acts to further both of these interests simultaneously, no ethical dilemma exists. In contrast, when either the interests or prior wishes of the pregnant patient diverge from the interests of the fetus, representatives may find themselves in a challenging bind. Across time and geography, different cultures have placed different degrees of weight on the value of the parent's life, and to a lesser degree the parent's health or preferences, vis-à-vis the fetus. At present, both American Constitutional law and the general consensus in Western bioethics strongly favors preserving parental life and health in the absence of clear indication that pregnant individual holds preferences to the contrary. However, in the absence of a serious threat to parental life or health, some jurisdictions, such as Texas, limit the authority of agents to endanger fetal welfare, or even temporarily suspend the authority of advance directives during pregnancy.[264] Such "pregnancy clauses" have generally been upheld by courts, and in the aftermath of the United States Supreme Court decision in *Dobbs v. Jackson Women's Health Organization* (2022), allowing states to criminalize elective abortions, that appears unlikely to change any time soon.[265]

Ethical and legal concerns about third-party decision-making for pregnant patients are magnified in situations in which the patient has suffered an illness or injury requiring life support. For example, should a pregnant patient's prior express wish to terminate life support, if that patient were ever to find herself in a particular medical condition, override the goal of preserving fetal life? And if the patient did not previously address the issue of pregnancy in particular, should the mere condition of being pregnant be treated as a variable that might have led the patient to a different preference? Even if an agent focuses solely upon effectuating the pregnant patient's autonomy, without any concern for fetal welfare, that agent might need to take into account how "the woman's prognosis, the viability of the fetus, the probable result of treatment and non-treatment for both mother and fetus and the mother's likely interest in avoiding impairment for her child together with her own instincts for survival, should be weighed against each other."[266] Yet most patients have likely never contemplated such complex factors in advance. Clinicians and agents may even confront circumstances in which a pregnant patient is legally brain dead – which then raises the question of to what degree the patient's prior wishes regarding parenthood or bringing the child to term should guide decisions to remove artificial ventilation or similar life support.[267] All of these issues defy easy consensus.

4.6.4 Defining the Default

In many cases where the law requires that an incapacitated patient's "best interests" be used to choose a course of care, the "best interests" of that individual are not entirely clear. How best to balance the goals of extending life and of reducing suffering, whether physical or existential, is a philosophical question rather than a scientific one. While the principle of autonomy creates room both for individuals who wish to have their lives preserved at any personal cost, as well as those who believe life beyond a certain point is undignified or not worth living, no easy method exists for determining which values should be prioritized as a default when a patient's wishes have never been voiced or cannot ever be known. Historically, Western bioethics and Anglo-American law have favored life extension as a default. Increasingly, some scholars have criticized this assumption. Many approaches exist to help guide society to answers in this regard. For instance, one could harness the democratic process and establish as a default the preference that would be voiced regarding care by a societal majority, noting that the issue of how to define the denominator is itself challenging. For example, for a patient born into an Amish family, would the majority consensus reflect that of the average American or the average

Mennonite? Alternatively, one might also consider the social and economic implications for others in determining whether or not to withdraw care as a default. What should be recognized is that no answer here is values-free, and depending upon one's own personal and cultural biases, one may construct a persuasive, if not dispositive, argument for a range of positions. In short, as in many areas of medical decision-making, there are no easy answers.

Notes

Capacity, Informed Consent, and Third-Party Decision-Making

1. Slack 1977.
2. Oken 1961.
3. Glaser 1965.
4. Howley 2018.
5. Does the doctor know best? – a professional opinion. 1975.
6. Wilkinson 1984.
7. Novack et al. 1979.
8. O'Connor v. Donaldson 1975.
9. Hacking 2004.
10. Szasz & Hollender 1956.
11. Szasz et al. 1958.
12. Hall et al. 2009.
13. Hall et al. 2009.
14. Ray 1877.
15. Renton 1890.
16. Green 1941.
17. Hess 1961.
18. Appelbaum 1986.
19. Appel 2023b.
20. Olin & Olin 1975.
21. Owens 1977.
22. Roth, Meisel & Lidz 1977.
23. Roth, Meisel & Lidz 1977.
24. Roth, Meisel & Lidz 1977.
25. Roth, Meisel & Lidz 1977.
26. Roth, Meisel & Lidz 1977.
27. Roth, Meisel & Lidz 1977.
28. Appelbaum & Roth 1982.
29. Drane 1985.
30. Drane 1985.
31. Drane 1985.
32. Drane 1985.
33. Drane 1985.
34. Herreros, Real de Asua & Palacios 2017.
35. Appel 2022d.
36. Culver & Gert 1990.
37. Wicclair 1991.
38. Buchanan & Brock 1986.
39. Buchanan & Brock 1986.
40. Buchanan & Brock 1986.

41. Buchanan & Brock 1986.
42. President's Commission 1982.
43. President's Commission 1982.
44. President's Commission 1982.
45. President's Commission 1982.
46. Appelbaum Grisso 1988.
47. Appelbaum Grisso 1988.
48. Appelbaum Grisso 1988.
49. Leykin, Roberts & Derubei 2011.
50. Lyons 2020.
51. Appelbaum Grisso 1988.
52. Appelbaum Grisso 1988.
53. Appelbaum Grisso 1988.
54. Family Health Care Decisions Act 2010.
55. Mirza & Appel 2023.
56. Mirza & Appel 2023.
57. Mirza & Appel 2023.
58. Mirza & Appel 2023.
59. Mirza & Appel 2023.
60. Mirza & Appel 2023.
61. Jones v. Star Credit Corp. 1969.
62. Morgan & Veitch 2004.
63. Flynn 2019.
64. Marson et al. 1997.
65. Kitamura 2000.
66. Washington 2019.
67. Willoughby 2018.
68. Shim 2021.
69. Schwartz & Blankenship 2014.
70. Garrett et al. 2023.
71. Schonfield 2008.
72. Kreitler 1999.
73. Appel 2022d.
74. Appel 2022d.
75. Appel 2022d.
76. Appel 2022d.
77. Idaho Code 39–4302 1977.
78. Uniform Health-Care Decisions Act 2016.
79. Appel 2023b.
80. Appel 2023b.
81. Appel 2023b.
82. Health Care Consent Act 1996.
83. Johnston & Liddle 2007.
84. Appelbaum & Grisso 1988.
85. Appel 2023a.
86. Hurst 2004.

87. Appel 2023a.
88. Appel 2023a.
89. Appel 2023a.
90. Appel 2023a.
91. Owen 2019.
92. Powderly 2000.
93. Cutter 2016.
94. Beauchamp 2011.
95. Cocanour 2017.
96. Mangold 1975.
97. American Medical Association 2016.
98. Union Pacific Railway Company v. Botsford 1891.
99. Union Pacific Railway Company v. Botsford 1991.
100. Union Pacific Railway Company v. Botsford 1991.
101. Mohr v. Williams 1905.
102. Mohr v. Williams 1905.
103. Keating 2021.
104. Mohr v. Williams 1905.
105. Pratt v. Davis 1906.
106. Horan & Halligan 1982.
107. Pratt v. Davis 1906.
108. Pratt v. Davis 1906.
109. Pratt v. Davis 1906.
110. Rolater v. Strain 1913.
111. Rolater v. Strain 1913.
112. Bazzano, Durant & Brantley 2021.
113. Schloendorff v. Society of New York Hospital 1914.
114. Chervenak, McCullough & Chervenak FA 2015.
115. Chervenak, McCullough & Chervenak FA 2015.
116. Chervenak, McCullough & Chervenak FA 2015.
117. Chervenak, McCullough & Chervenak FA 2015.
118. Schloendorff v. Society of New York Hospital 1914.
119. Lombardo 2005.
120. Bing v. Thunig 1957.
121. United States of America v. Karl Brandt et al. 1947.
122. Kurtz 2020.
123. Hunt v. Bradshaw 1955.
124. Salgo v. Leland Stanford Jr. University Board of Trustees 1957.
125. Salgo v. Leland Stanford Jr. University Board of Trustees 1957.
126. Salgo v. Leland Stanford Jr. University Board of Trustees 1957.
127. Mitchell v. Robinson 1960.
128. Cobbs v. Grant 8 1972.
129. Churchward 1990.
130. Churchward 1990.
131. Churchward 1990.
132. Small v Howard 1880.

133. Small v Howard 1880.
134. Lewis, Gohagan & Merenstein 2007.
135. Robinson v LeCorps 2000.
136. Brune v. Belinkoff 1968.
137. Cooke, Worsham & Reisfield 2017.
138. Waltz 1969.
139. T. J. Hooper 1932.
140. Helling v. Carey 1974.
141. Nelson 2002.
142. Helling v. Carey 1974.
143. Helling v. Carey 1974.
144. Canterbury v. Spence 1972.
145. Canterbury v. Spence 1972.
146. Canterbury v. Spence 1972.
147. Murphy 1976.
148. Sidaway v Bethlem Hospital 1985.
149. Montgomery v Lanarkshire Health Board 2015.
150. Appelbaum 1997.
151. Osheroff v. Chestnut Lodge, Inc. 1985.
152. Klerman 1990.
153. Klerman 1990.
154. Bernat 2005.
155. Angelucci 2007.
156. Brown 1993.
157. Kapley et al. 2013.
158. O'Neill 2003.
159. Bester, Cole & Kodish 2016.
160. Plaut 1989.
161. Hinkle 1981.
162. Shalak, Shariff, Doddapaneni & Suleman 2022.
163. Somerville 1984.
164. Somerville 1984.
165. Shalak, Shariff, Doddapaneni & Suleman 2022.
166. Menon, Entwistle, Campbell & van Delden 2021.
167. New South Wales Health 2020.
168. Edwin 2008.
169. Edwin 2008.
170. Bonomo, Rinderknecht & Beiser 2003.
171. Appel 2004.
172. Kutner 1979.
173. Williams 1957.
174. Kutner 1969.
175. Kutner 1969.
176. Kutner 1979.
177. Kutner 1969.
178. Kutner 1979.

179. Miller 2017.
180. Miller 2017.
181. In Re Quinlan 1976.
182. Jonsen 1978.
183. Annas 1991.
184. Teoli & Ghassemzadeh 2023.
185. Annas 1991.
186. Lyden 2006.
187. Stern 1993.
188. Stern 1993.
189. Mathew, Gershengorn & Hua 2018.
190. 8 V.S.A. Ch. 231 § 9701(4)(A) 2011.
191. Moye, Sabatino & Weintraub 2013.
192. Appel 2022a.
193. Howe 2014.
194. Howe 2014.
195. Reamy, Kim, Zarit & Whitlatch 2011.
196. Seckler & Meier & Mulvihill & Paris 1991.
197. Wolinsky et al. 2016.
198. Seckler, Meier, Mulvihill & Paris 1991.
199. Sulmasy, Haller & Terry 1994.
200. Caralis, Davis, Wright & Marcial 1993.
201. Morrison et al. 1998.
202. Thomas 2009.
203. Shea 2021.
204. O'Brien & Fiester 2014.
205. Zuger 2017.
206. Courtwright & Rubin 2016.
207. Pope 2017.
208. Farrell et al. 2017.
209. Courtwright & Rubin 2016.
210. Farrell et al. 2017.
211. Torke 2008.
212. Buchanan & Brock 1990.
213. Gutheil & Appelbaum 1983.
214. Ex parte Whitbread in re Hinde, a Lunatic(1816); Strunk v. Strunk 1969.
215. Wendler & Phillips 2015.
216. Gutheil & Appelbaum 1983.
217. Appel 2024.
218. Appel 2022c.
219. Dworkin 1994.
220. Combs, Rasinski, Yoon & Curlin 2013.
221. Broström, Johansson & Nielsen 2007.
222. Dresser & Robertson 1989.
223. Paris et al. 2017.
224. Lee et al. 2006.

225. Wasserman, Navin &Vercler 2019.
226. Shapiro 2012.
227. Meadow 1999.
228. Lando 2002.
229. Seltzer 2016.
230. Cruzan v. Director, Missouri Department of Health 1990.
231. Annas 1991.
232. Cruzan v. Harmon 1988.
233. Cruzan v. Director, Missouri Department of Health 1990.
234. Wolf 1990.
235. Calabresi 2006.
236. Lewin 1990.
237. Rennie v. Klein 1978.
238. Rogers v. Commissioner of Mental Health 1983.
239. Cuca 1992–1993.
240. Cuca 1992–1993.
241. Murray & Wortzel 2019.
242. Appelbaum 2004.
243. Appelbaum 2004.
244. Brenna et al. 2023.
245. Brenna et al. 2023.
246. Gaillard et al. 2023.
247. Potiki et al. 2023.
248. Appel 2021c.
249. Appel 2021.
250. Appel 2021.
251. Appel 2007.
252. van Veen, Widdershoven, Beekman & Evans 2022.
253. Superintendent of Belchertown State School v. Saikewicz 1977.
254. Superintendent of Belchertown State School v. Saikewicz 1977.
255. Lee 1980.
256. Matter of Storar 1981.
257. Matter of Storar 1981.
258. Buchanan 1979.
259. Appel 2022b.
260. Appel 2022b.
261. Appel 2022b.
262. Appel 2022e.
263. Appel 2022.
264. Taylor 2014.
265. Dyke 1990.
266. Lemmens 2010.
267. Sperling 2020.

Bibliography

8 V.S.A. Ch. 231 § 9701(4)(A) (2011).

American Medical Association. Withholding Information from Patients. *AMA Principles of Medical Ethics* Opinion 8.122 (2016). https://code-medical-ethics.ama-assn.org/principles.

Angelucci, Patricia A. What Is Medical Futility? *Nursing Critical Care* 2 (2007): 20–21.

Annas, George J. The Health Care Proxy and the Living Will. *New England Journal of Medicine* 325 (1991): 1210–13.

Appel, Jacob M. A Duty to Kill? A Duty to Die? Rethinking the Euthanasia Controversy of 1906. *Bulletin of the History of Medicine* 78 (2004): 610–34.

Appel, Jacob M. A Suicide Right for the Mentally Ill. *The Hastings Center Report* 37 (2007): 21–23.

Appel, Jacob M. Capital Punishment, Psychiatrists and the Potential Bottleneck of Competence. *Journal of Law & Health* 24 (2011): 45–78.

Appel, Jacob M. Reconsidering Capacity to Appoint a Healthcare Proxy. *Cambridge Quarterly of Healthcare Ethics* 32 (2022a): 1–7.

Appel, Jacob M. Substituted Judgment for the Never-Capacitated: Crossing Storar's Bridge Too Far. *Bioethics* 36 (2022b): 225–31.

Appel, Jacob M. Trial by Triad: Substituted Judgment, Mental Illness and the Right to Die. *Journal of Medical Ethics* 48 (2022c): 358–61.

Appel, Jacob M. A Values-Based Approach to Capacity Assessment. *Journal of Legal Medicine* 42 (2022d): 53–65.

Appel, Jacob M. Locating and Identifying Third-Party Decision-Makers. *Journal of the American Academy of Psychiatry & Law* 50 (2022e): 84–96.

Appel, Jacob M. Anything You Do Not Say Can Be Used against You: Volitional Refusal to Engage in Decisional Capacity Assessment. *Journal of Clinical Ethics* 34 (2023a): 204–10.

Appel, Jacob M. The Statutory Codification of Decisional Capacity Standards. *Journal of the American Academy of Psychiatry & Law* 51 (2023b): 230043–23.

Appel, Jacob M. Anything You Do Not Say Can Be Used against You: Volitional Refusal to Engage in Decisional Capacity Assessment. *Journal of Clinical Ethics* 34 (2023c): 204–10.

Appel Jacob M. Decisional Capacity After Dark: Is Autonomy Delayed Truly Autonomy Denied?. *Cambridge Quarterly of Healthcare Ethics* 33 (2024): 260–66.

Appelbaum, Paul S. Competence to Be Executed: Another Conundrum for Mental Health Professionals. *Psychiatric Services* 37 (1986): 682–84.

Appelbaum, Paul S. Law & Psychiatry: Psychiatric Advance Directives and the Treatment of Committed Patients. *Psychiatric Services* 55 (2004): 751–52.

Appelbaum, Paul S. Informed Consent to Psychotherapy: Recent Developments. *Psychiatric Services* 48 (1997): 445–46.

Appelbaum, Paul S. Law & Psychiatry: Psychiatric Advance Directives and the Treatment of Committed Patients. *Psychiatric Services* 55 (2004): 751–52, 763.

Appelbaum, Paul S. & Roth, Loren H. Competency to Consent to Research: A Psychiatric Overview. *Archives of General Psychiatry* 39 (1982): 951–58.

Appelbaum, Paul S. & Grisso, Thomas. Assessing Patients' Capacities to Consent to Treatment. *New England Journal of Medicine* 319 (1988): 1635–38.

Armontrout, James, Gitlin, David & Gutheil, Thomas. Do Consultation Psychiatrists, Forensic Psychiatrists, Psychiatry Trainees, and Health Care Lawyers Differ in Opinion on Gray Area Decision-Making Capacity Cases? A Vignette-Based Survey. *Psychosomatics* 57 (2016): 472–79.

Bazzano, Lydia A., Durant, Jaquail & Brantley, Paula Rhode. A Modern History of Informed Consent and the Role of Key Information. *Ochsner Journal* 21 (2021): 81–85.

Beauchamp, Tom L. Informed Consent: Its History, Meaning, and Present Challenges. *Cambridge Quarterly of Healthcare Ethics* 20 (2011): 515–23.

Bernat, James L. Medical Futility: Definition, Determination, and Disputes in Critical Care. *Neurocritical Care* 2 (2005): 198–205.

Bester, Johan, Cole, Christie M. & Kodish, Eric. The Limits of Informed Consent for an Overwhelmed Patient: Clinicians' Role in Protecting Patients and Preventing Overwhelm. *AMA Journal of Ethics* 18 (2016): 869–886.

Bing v. Thunig, 2 N.Y.2d 656 (1957).

Bonomo, Jordan, Rinderknecht, Tanya N. & Beiser, Edward N. "It Ain't Easy Being Green," A Case-Based Analysis of Ethics and Medical Education on the Wards. *Medicine $ Health, R. I.* 86 (2003): 276–78.

Braun, Michelle, Ronald Gurrera, Michele Karel, Jorge Armesto & Jennifer Moye. Are Clinicians Ever Biased in Their Judgments of the Capacity of Older Adults to Make Medical Decisions. *Generations* 33 (2009): 78–91.

Brenna, Connor T. A., Stacy S Chen, Matthew Cho, Liam G McCoym & Sunit Das. Steering Clear of Akrasia: An Integrative Review of Self-Binding Ulysses Contracts in Clinical Practice. *Bioethics* 37 (2023): 690–714.

Broström, Linus, Johansson, Mats & Nielsen, Morten K. "What the Patient Would Have Decided": A Fundamental Problem with the Substituted

Judgment Standard. *Medicine, Health Care and Philosophy* 10 (2007): 265–78.

Brown, Douglas R. Panacea or Pandora's Box: The "Two Schools of Medical Thought" Doctrine after Jones v. Chidester, 610 A.2d 964 (Pa. 992). *Washington University Journal of Urban and Contemporary Law* 44 (1993): 223–34.

Brune v. Belinkoff, 235 N.E.2d 793 (Mass. 1968).

Buchanan, Allen. Medical Paternalism or Legal Imperialism: Not the Only Alternatives for Handling Saikewicz-Type Cases. *American Journal of Law and Medicine* 5 (1979): 97–117.

Buchanan, Allen & Brock, Dan W. Deciding for Others. *Milbank Quarterly* 64 (1986): 17–94.

Buchanan, Allen & Brock, Dan W. *Deciding for Others*. Cambridge: Cambridge University Press, 1990.

Calabresi, Steven G. The Terri Schiavo Case: In Defense of the Special Law Enacted by Congress and President Bush. *Northwestern University Law Review* 100 (2006): 164–66.

Canterbury v. Spence (464 F.2d. 772) (1972).

Caralis, Panagiota V., Davis, Bobbi, Wright, Karen & Marcial, Eileen. The Influence of Ethnicity and Race on Attitudes toward Advance Directives, Life-Prolonging Treatments, and Euthanasia. *Journal of Clinical Ethics* 4 (1993): 155–65.

Chervenak, Judith, McCullough, Laurence B. & Chervenak, Frank A. Surgery without Consent or Miscommunication? A New Look at a Landmark Legal Case. *America Journal of Obstetrics and Gynecology* 212 (2015): 586–90.

Churchward, Adrian B. A Comparative Study of the Law Relating to the Physician's Duty to Obtain the Patient's "Informed Consent" to Medical Treatment in England and California. *Connecticut Journal of International Law* 5 (1990): 483–563.

Cobbs v. Grant, 8 Cal 3d 229 (1972).

Cocanour, Christine S. Informed Consent – It's More than a Signature on a Piece of Paper. *American Journal of Surgery* 214 (2017): 993–97.

Combs, Michael P., Rasinski, Kenneth A., Yoon, John D. & Curlin, Farr A. Substituted Judgment in Principle and Practice: A National Physician Survey. *Mayo Clinic Proceedings* 88 (2013): 666–73.

Cooke, Brian K., Worsham, Elizabeth & Reisfield, Gary M. The Elusive Standard of Care. *Journal of the American Academy of Psychiatry & Law* 45 (2017): 358–64.

Courtwright, Andrew & Rubin, Emily. Who Should Decide for the Unrepresented? *Bioethics* 30 (2016): 173–80.

Cruzan v. Director, Missouri Department of Health, 497 U.S. 261 (1990).

Cruzan v. Harmon, 760 S.W. 2d 408, 416417 (1988).

Cuca, Roberto. Ulysses in Minnesota: First Steps toward a Self-Binding Psychiatric Advance Directive Statute. *Cornell Law Review* 78 (1992–1993): 1152–86.

Culver, Charles M. & Gert, Bernard. The Inadequacy of Incompetence. *Milbank Quarterly* 68 (1990): 619–43.

Cutter, Laura. Walter Reed, Yellow Fever, and Informed Consent. *Military Medicine* 181 (2016): 90–91.

Davidson, Henry A. Testimonial Capacity. *Boston University Law Review* 39 (1959): 172–80.

Does the Doctor Know Best? – A Professional Opinion. *The Listener* (London), December 18, 1975.

Drane, James F. The Many Faces of Competency. *Hastings Center Report* 15 (1985): 17–21.

Dresser, Rebecca S. & Robertson, John A. Quality of Life and Non-Treatment Decisions for Incompetent Patients: A Critique of the Orthodox Approach. *Law Medicine & Health Care* 17 (1989): 234–44.

Dworkin, Ronald. *Life's Dominion: An Argument about Abortion, Euthanasia, and Individual Freedom.* New York: Vintage Books, 1994.

Dyke, Molly C. A Matter of Life and Death: Pregnancy Clauses in Living Will Statutes. *Boston University Law Review* 70 (1990): 867–87.

Edwin, A. K. Don't Lie but Don't Tell the Whole Truth: The Therapeutic Privilege – Is It Ever Justified? *Ghana Medical Journal* 42 (2008): 156–61.

Ex parte Whitbread in re Hinde, a Lunatic, 35 Eng. Rep. 878 (1816); Strunk v. Strunk, 445 S.W.2d 145 (Ky. Ct. App. 1969).

Family Health Care Decisions Act, Public Health Law Ch. 29-CC, (New York, 2010).

Farrell, Timothy W, Widera Eric, Rosenberg, Lisa et al. Clinical Practice and Models of Care, and Public Policy Committees of the American Geriatrics Society. AGS Position Statement: Making Medical Treatment Decisions for Unbefriended Older Adults. *Journal of the American Geriatric Society* 65 (2017): 14–15.

Flynn, Eilionóir. The Rejection of Capacity Assessments in Favor of Respect for Will and Preferences: The Radical Promise of the UN Convention on the Rights of Persons with Disabilities. *World Psychiatry* 18 (2019): 50–51.

Gaillard, Anne-Sophie, Braun, Esther, Vollmann, Jochen, Gather, Jakov & Scholten Matthé. The Content of Psychiatric Advance Directives: A Systematic Review. *Psychiatric Services* 74 (2023): 44–55.

Garrett W. S., Verma A., Thomas D., Appel J. M., Mirza O. Racial Disparities in Psychiatric Decisional Capacity Consultations. *Psychiatric Services* 74 (2023):10–16.

Glaser, Barney & Strauss, Anselm. *Awareness of Dying*. London: Weindenfeld and Nicolson, 1965.

Green, Milton D. Judicial Tests of Mental Incompetency. *Missouri Law Review* 6 (1941): 141–65.

Gutheil, Thomas G. & Appelbaum, Paul S. Substituted Judgment: Best Interests in Disguise. *Hastings Center Report* 13 (1983): 8–11.

Hacking, Ian. Between Michel Foucault and Erving Goffman: Between Discourse in the Abstract and Face-To-Face Interaction. *Economy and Society* 33 (2004): 277–302.

Hall, Ryan C.W., Richard, C.W. Hall, Myers, W. C. & Chapman, M. J. Testamentary Capacity: History, Physicians' Role, Requirements, and Why Wills Are Challenged. *Clinical Geriatrics* 17 (2009): 18–24.

Health Care Consent Act, S.O. 1996, c. 2, Sched. A (1996).

Helling v. Carey, 83 Wn.2d 514, 519 P.2d 981 (1974).

Hermann, Helena, Trachsel, Manuel & Biller-Andorno. Nikola: Physicians' Personal Values in Determining Medical Decision-Making Capacity: A Survey Study. *Journal of Medical Ethics* 41 (2015): 739–44.

Herreros, Benjamín, Real de Asua, Diego & Palacios, Gregorio. Are We Still Our Patients' Keepers?: James Drane's Contribution to Clinical Ethics in the Current Context. *Journal of Healthcare Quality Research* 33 (2017): 54–59.

Hess, John H., Pearsall, Henry B., Slichter, Donald & Thomas, Herbert E. Criminal Law: Insane Persons: Competency to Stand Trial. *Michigan Law Review* 59 (1961): 1078–1100.

Hinkle, Bruce J. Informed Consent and the Family Physician. *Journal of Family Practice* 12 (1981): 109–15.

Horan, Dennis J. & Halligan, Patrick. Informed Consent. *Linacre Quarterly* 49 (1982): 358–69.

Howe, Edmund G. New Approaches with Surrogate Decision Makers. *Journal of Clinical Ethics* 25 (2014): 261–72.

Howley, Elaine K. From the "Big C" to "Cancer": Cultural Taboos Surrounding the Disease Have Lessened, but Continued Research and Education Are Still Needed. *US News & World Reports*, March 21, 2018.

Hunt v. Bradshaw, 88 S.E.2d 762 (1955).

Hurst, Samia A. When Patients Refuse Assessment of Decision-Making Capacity: How Should Clinicians Respond? *Archives of Internal Medicine* 164 (2004): 1757–60.

Idaho Code 39–4302 (1977).

In Re Quinlan, 70 N.J. 10 (1976).

Johnston, Carolyn & Liddle, Jane. The Mental Capacity Act 2005: A New Framework for Healthcare Decision Making. *Journal of Medical Ethics* 33 (2007): 94–97.

Jones v. Star Credit Corp. 59 Misc. 2d 189 (1969).

Jonsen, Albert R. Dying Right in California–The Natural Death Act. *Clinical Toxicology* 13 (1978): 513–22.

Kapley, David, Appel, Jacob M., Resnick, P. J. Mental Health Innovation vs. Psychiatric Malpractice: Creating Space for "Reasonable Innovation." *Faulkner Law Review* 5 (2013).

Keating, Gregory C. The Theory of Enterprise Liability and Common Law Strict Liability. *Vanderbilt Law Review* 54 (2001): 131–64.

Kitamura, Toshinori. Assessment of Psychiatric Patients' Competency to Give Informed Consent: Legal Safeguard of Civil Right to Autonomous Decision-Making. *Psychiatry and Clinical Neurosciences* 54 (2000): 515–22.

Klerman, Gerald L. The Psychiatric Patient's Right to Effective Treatment: Implications of Osheroff V. Chestnut Lodge. *American Journal of Psychiatry* 147 (1990): 409–18.

Kreitler, Shulamith. Denial in Cancer Patients. *Cancer Investigation* 17 (1999): 514–34.

Kurtz, Sheldon F. The Law of Informed Consent: From "Doctor Is Right" to "Patient Has Rights. *Syracuse Law Review* 50 (2000): 1243–60.

Kutner, Luis. Due Process of Euthanasia: The Living Will, a Proposal. *Indiana Law Journal* 44 (1969): 539–54.

Kutner, Luis. Euthanasia: Due Process for Death with Dignity; The Living Will. *Indiana Law Journal* 54 (1979): 201–28.

Lando, Henrik. When Is the Preponderance of the Evidence Standard Optimal? *The Geneva Papers on Risk and Insurance* 27 (2002): 602–608.

Lee, Daniel E. The Saikewicz Decision and Patient Autonomy. *Linacre Quarterly* 47 (1980): 64–69.

Lee, K. Jane, Havens, Peter L., Sato, Thomas T., Hoffman, George M. & Leuthner, Steven R. Assent for Treatment: Clinician Knowledge, Attitudes, and Practice. *Pediatrics* 118 (2006): 723–30.

Lemmens, Christophe. End of Life Decisions and Pregnant Women: Do Pregnant Women Have the Right to Refuse Life Preserving Medical Treatment? A Comparative Study. *European Journal of Health Law* 17 (2010): 485–505.

Lewin, Tamar. Nancy Cruzan Dies, Outlived by a Debate over the Right to Die. *The New York Times*. December 27, 1990.

Lewis, Michelle H., Gohagan, John K. & Merenstein, Daniel J. The Locality Rule and the Physician's Dilemma: Local Medical Practices vs. the National Standard of Care. *JAMA*. June 20, 2007; 297(23): 2633–37.

Leykin, Yan, Roberts, Carolyn Sewell & Derubei, Robert J. Decision-Making and Depressive Symptomatology. *Cognitive Therapy and Research* 35 (2011): 333–41.

Lombardo, Paul A. Phantom Tumors and Hysterical Women: Revising Our View of the Schloendorff Case. *Journal of Law, Medicine & Ethics* 33 (2005): 791–801.

Lyden, Martin. Capacity Issues Related to the Health Care Proxy. *Mental Retardation* 44 (2006): 272–82.

Lyons, Siobhan. Do We Want to Be Free? *Philosophy Now* 8 (2020): 22–24.

Mangold, William J. Jr. Informed Consent and Its Implications in Family Practice. *Journal of Family Practice* 2 (1975): 103–105.

Marson, Daniel C., McInturff, Bronwyn, Hawkins, Lauren, Bartolucci, Alfred & Harrell, Lindy E. Consistency of Physician Judgments of Capacity to Consent in Mild Alzheimer's Disease. *Journal of the American Geriatrics Society* 45 (1997): 453–57.

Mathew, Sharon, Gershengorn, Hayley B. & Hua, May. Terminology for Surrogate Decision Making Varies Widely by State. *Journal of Palliative Medicine* 21 (2018): 1060–61.

Matter of Storar, 52 N.Y.2d 363 (1981).

Mccurdy, William Edward. Insanity as a Ground for Annulment or Divorce in English and American Law. *American Journal of Psychiatry* 100 (1943): 185–201.

Meadow, Robin. Clear and Convincing Evidence: How Much Is Enough? *California Insurance Litigator* (1999): 116–21.

Meisel, Alan, Roth, Loren H. & Lidz, Charles W. Toward a Model of the Legal Doctrine of Informed Consent. *American Journal of Psychiatry* 134 (1977): 285–89.

Menon, Sumytra, Entwistle, Vikki A., Campbell, Alastair V. & van Delden, Johannes J. M. How Should the "Privilege" in Therapeutic Privilege Be Conceived When Considering the Decision-Making Process for Patients with Borderline Capacity? *Journal of Medical Ethics* 47 (2021): 47–50.

Miller, Blanca. Nurses in the Know: The History and Future of Advance Directives. *OJIN: The Online Journal of Issues in Nursing* 22 (2017).

Mirza, Omar F. & Appel, Jacob M. Capacity Reconceptualized: From Assessment Tool to Clinical Intervention. *Cambridge Quarterly of Healthcare Ethics* 24 (2023): 1–5.

Mitchell v. Robinson 334 S.W.2d 11 (1960).

Mohr v. Williams, 104 N.W. 12 (Minn. 1905).

Montgomery v Lanarkshire Health Board, UKSC 11 (2015).

Morgan, Derek & Veitch, Kenneth. Being Ms B: B, Autonomy and the Nature of Legal Regulation. *Sydney Law Review* 26 (2004): 107–30.

Morrison, R. Sean, Zayas, Luis H., Mulvihill, Michael, Baskin, Shari A. & Meier, Diane E. Barriers to Completion of Health Care Proxies: An Examination of Ethnic Differences. *Archives of Internal Medicine* 158 (1998): 2493–97.

Moye, Jennifer, Sabatino, Charles P. & Weintraub Brendel, Rebecca. Evaluation of the Capacity to Appoint a Healthcare Proxy. *American Journal of Geriatric Psychiatry* 21 (2013): 326–36.

Murphy, Walter J. Canterbury v. Spence–the Case and a Few Comments. *Forum.* 1976 Spring; 11(3): 716–26.

Murray, Heather & Wortzel, Hal. Psychiatric Advance Directives: Origins, Benefits, Challenges, and Future Directions. *Journal of Psychiatric Practice*, 2019; 25 (4): 303–307.

Natanson v. Klein 187 Kan. 186 (Kan. 1960).

Nelson, Leonard J. Helling V. Carey Revisited: Physician Liability in the Age of Managed Care (March 23, 2010). *Seattle University Law Review* 25 (2002): 775–819.

New South Wales Health. Consent to Medical and Healthcare Treatment Manual. (2020). www.health.nsw.gov.au/policies/manuals/Documents/consent-section-4.pdf.

Novack, Dennis H., Plumer, Robin, Smith, Raymond L. et al. Changes in Physicians' Attitudes toward Telling the Cancer Patient. *JAMA* 241 (1979): 897–900.

O'Brien, Meghan & Fiester, Autumn. Who's at the Table? Moral Obligations to Equal-Priority Surrogates in Clinical Ethics Consultations. *Journal of Clinical Ethics* 25 (2014): 273–80.

O'Connor v. Donaldson, 422 U.S. 563 (1975).

O'Neill O. Some Limits of Informed Consent. *Journal of Medical Ethics* 29 (2003): 4–7.

Oken, Donald. What to Tell Cancer Patients: A Study of Medical Attitudes. *JAMA* 175 (1961): 1120–28.

Olin, Grace B. & Olin, Harry S. Informed Consent in Voluntary Mental Hospital Admissions. *American Journal of Psychiatry* 132 (1975): 938–41.

Osheroff v. Chestnut Lodge, Inc., 490 A. 2d 720 (1985).

Owen, Adrian M. The Search for Consciousness. *Neuron* 102 (2019): 526–28.

Owens, Howard. When Is a Voluntary Commitment Really Voluntary? *American Journal of Orthopsychiatry* 47 (1977): 104–10.

Paris, John J., Ahluwalia, Jag, Cummingsm, Brian, Moreland Michael P., Wilkinson, Dominic. The Charlie Gard Case: British and American Approaches to Court Resolution of Disputes over Medical Decisions. *Journal of Perinatology* 37 (2017): 1268–71.

Plaut, Eric A. The Ethics of Informed Consent: An Overview. *Psychiatric Journal of the University of Ottawa* 14 (1989): 435–38.

Pope, Thaddeus. Unbefriended and Unrepresented: Better Medical Decision Making for Incapacitated Patients without Healthcare Surrogates. *Georgia State University Law Review* 33 (2017): 923–1020.

Potiki, Johnnie, Tawaroa, Daniel, Casey, Heather et al. Cultural Influences on the Creation and Use of Psychiatric Advance Directives. *Psychiatric Services* 74 (2023): 1299–302.

Powderly, Kahtleen E. Patient Consent and Negotiation in the Brooklyn Gynecological Practice of Alexander J.C. Skene: 1863–1900. *Journal of Medicine and Philosophy.* February 2000; 25(1): 12–27.

Pratt v. Davis, 224 Ill. 300, 79 N.E. 562 (1906).

President's Commission for the Study of Ethical Problems in Medicine Biomedical Behavioral Research. Making Health Care Decisions: The Ethical and Legal Implications of Informed Consent in the Patient-Practitioner Relationship. President's Commission for the Study of (1982).

Ray, Isaac. Testamentary Capacity. *The Sanitarian* 5 (1877): 433–46.

Reamy, Allison M., Kim, Kyungmin, Zarit, Steven H. & Whitlatch, Carol J. Understanding Discrepancy in Perceptions of Values: Individuals with Mild to Moderate Dementia and Their Family Caregivers. *The Gerontologist* 51 (2011): 473–83.

Rennie v. Klein, 462 F. Supp. 1131 (D.N.J. 1978).

Renton, A. Wood. Insanity in Its Relation to Contract. *Cape Law Journal* 5 (1888): 117–20.

Renton, A. Wood. The Legal Test of Lunacy. *Medico-Legal Journal* 8 (1890): 317–19.

Robinson v LeCorps, 83 SW3d 718 (Tenn. 2002).

Rogers v. Commissioner of Mental Health, 390 Mass.489, 458 N.E.2d 308. (1983).

Rolater v. Strain, 39 Okla. 572, 137 P. 86 (1913).

Roth, William F. The Role of the Psychiatrist under the M'Naghten Rule. *University of Kansas City Law Review* 28 (1959): 133–38.

Roth, Loren H., Meisel, Alan & Lidz, Charles W. Tests of Competency to Consent to Treatment. *American Journal of Psychiatry* 134 (1977): 279–84.

Salgo v. Leland Stanford Jr. University Board of Trustees, 154 Cal.App.2d 560, 317 P.2d 170 (1957).

Schloendorff v. Society of New York Hospital, 105 N.E. 92 (N.Y. 1914).

Schonfield, Toby L. *Ethics by Committee: A Textbook on Consultation, Organization, and Education for Hospital Ethics Committees* (ed. Hester, Micah D.) Lanham, MD: Rowman & Littlefield, 2008.

Schwartz, Robert C. & Blankenship, David M. Racial Disparities in Psychotic Disorder Diagnosis: A Review of Empirical Literature. *World Journal of Psychiatry* 22 (2014): 133–40.

Seckler, Allison B., Meier, Diane E., Mulvihill, Michael & Paris, Barbara E. Cammer. Substituted Judgment: How Accurate Are Proxy Predictions? *Annals of Internal Medicine* 115 (1991): 92–98.

Seltzer, Richard, Hansberry, Heidi L., Canan, Russell F., Cannon, Molly & Seltzer, Richard. Legal Standards by the Numbers: Quantifying Burdens of Proof or a Search for Fool's Gold. *Judicature* 100 (2016): 56.

Shalak, Manar, Shariff, Masood A., Doddapaneni, Varun & Suleman, Natasha. The Truth, the Whole Truth, and Nothing but the Truth: Therapeutic Privilege. *Journal Postgraduate Medicine* 68 (2022): 152–55.

Shapiro, Barbara J. *Beyond Reasonable Doubt: The Evolution of a Concept: Fictions of Knowledge: Fact, Evidence, Doubt*. London: Palgrave Macmillan, 2012.

Shea, Matthew. The Ethics of Choosing a Surrogate Decision Maker When Equal-Priority Surrogates Disagree. *Narrative Inquiry in Bioethics* 11 (2021): 121–31.

Shim, Ruth S. Dismantling Structural Racism in Psychiatry: A Path to Mental Health Equity. *American Journal of Psychiatry* 178 (2021): 592–98.

Sidaway v Bethlem Hospital AC 871 (1985).

Slack, Warner V. The Patient's Right to Decide. *Lancet* 2 (1977): 240.

Small v Howard, 128 Mass 131 (1880).

Somerville, Margaret A. Therapeutic Privilege: Variation on the Theme of Informed Consent. *Law Medicine & Health Care* 12 (1984): 4–12.

Sperling, Daniel. Brain Dead Receive Life Support, Despite Objection from Her Appointed. *AMA Journal of Ethics* 22 (2020): E1004–E1009.

Stern, Kristina. Living Wills in English Law. *Palliative Medicine* 7 (1993): 283–88.

Strunk v. Strunk, 445 S.W.2d 145 (Ky. Ct. App. 1969).

Sulmasy, Daniel P., Haller, Karen & Terry, Peter B. More Talk, Less Paper: Predicting the Accuracy of Substituted Judgments. *American Journal of Medicine* 96 (1994): 432–38.

Superintendent of Belchertown State School v. Saikewicz 373 Mass. 728 (1977).

Szasz, Thomas S. & Hollender, Marc H. A Contribution to the Philosophy of Medicine: The Basic Models of the Doctor-Patient Relationship. *AMA Archives of Internal Medicine* 97 (1956): 585–92.

Szasz, Thomas S., Knoff, William F. & Hollender, Marc. The Doctor-Patient Relationship and Its Historical Context. *American Journal of Psychiatry* 115 (1958): 522–28.

T. J. Hooper, 60 F.2d 737 (2d Cir. 1932).

Taylor, Katherine. The Pregnancy Exclusions: Respect for Women Requires Repeal. *American Journal of Bioethics* 14 (2014): 50–52.

Teoli, Dac & Ghassemzadeh, Sassan. *Patient Self-Determination Act in StatPearls.* Treasure Island, FL: StatPearls, 2023.

Thomas, Jennifer. Common Law Marriage. *Journal of the American Academy of Matrimonial Law* 22 (2009): 151–67.

Torke, Alexia M., Alexander, G. Caleb & Lantos, John. Substituted Judgment: The Limitations of Autonomy in Surrogate Decision Making. *Journal of General Internal Medicine* 23 (2008): 1514–17.

Udelsman, Brooks V., Govea, Nicolas, Cooper, Zara et al. Concordance in Advance Care Preferences among High-Risk Surgical Patients and Surrogate Health Care Decision Makers in the Perioperative Setting. *Surgery* 167 (2020): 396–403.

Uniform Health-Care Decisions Act National Conference of Commissioners on Uniform State Laws. *Issues in Law & Medicine* 22 (2016): 83–97.

Union Pacific Railway Company v. Botsford, 141 U.S. 250 (1891).

United States of America v. Karl Brandt, et al. (1947).

Van Veen, Sisco, Widdershoven, Guy, Beekman, Aartjan & Evans Natalie. Physician Assisted Death for Psychiatric Suffering: Experiences in the Netherlands. *Frontiers in Psychiatry* 13 (2022): 895387.

Waltz, Jon R. The Rise and Gradual Fall of the Locality Rule in Medical Malpractice Litigation. *DePaul Law Review* 18 (1969): 408–20.

Washington, Harriet A. *Medical Apartheid.* New York: Random House, 2019.

Wasserman, Jason Adam, Navin, Mark Christopher & Vercler, Chrisian J. Pediatric Assent and Treating Children over Objection. *Pediatrics* 144 (2019): 7–9.

Wendler, David S. & Phillips, John. Clarifying Substituted Judgment: The Endorsed Life Approach. *Journal of Medical Ethics* 41 (2015): 723–30.

Wicclair, Marc R. Patient Decision-Making Capacity and Risk. *Bioethics* 5 (1991): 91–104.

Wilcox, Arthur W. Insanity and Marriage. *Westminster Review* 158 (1902): 199–209.

Wilkinson, James. Does Doctor Know Best? *The Listener* (London), June 21, 1984.

Williams, Glanville L. *The Sanctity of Life and the Criminal Law.* New York: Alfred A. Knopf, 1957.

Willoughby, Christopher. Running Away from Drapetomania: Samuel A. Cartwright, Medicine, and Race in the Antebellum South. *Journal of Southern History* 84 (2018): 579–614.

Wolf, Susan M. Nancy Beth Cruzan: In No Voice at All. *Hastings Center Report* 21 (1990): 38–41.

Wolinsky, Fredric D., Ayres, Lioness, Jones, Michael P. et al. A Pilot Study among Older Adults of the Concordance between Their Self-Reports to a Health Survey and Spousal Proxy Reports on Their Behalf. *BMC Health Services Research* 16 (2016): 485.

Zuger, Abigail. Legalities of Surrogate Medical Decision Making Vary by State. *NEJM Journal Watch. General Medicine* (2017).

Cambridge Elements ≡

Bioethics and Neuroethics

Thomasine Kushner

California Pacific Medical Center, San Francisco
Thomasine Kushner, PhD, is the founding Editor of the *Cambridge Quarterly of Healthcare Ethics* and coordinates the International Bioethics Retreat, where bioethicists share their current research projects, the Cambridge Consortium for Bioethics Education, a growing network of global bioethics educators, and the Cambridge-ICM Neuroethics Network, which provides a setting for leading brain scientists and ethicists to learn from each other.

About the Series

Bioethics and neuroethics play pivotal roles in today's debates in philosophy, science, law, and health policy. With the rapid growth of scientific and technological advances, their importance will only increase. This series provides focused and comprehensive coverage in both disciplines consisting of foundational topics, current subjects under discussion and views toward future developments.

Cambridge Elements $^{\equiv}$

Bioethics and Neuroethics

Printed in the United States
by Baker & Taylor Publisher Services